AF709028

AGRICULTURE ISSUES AND POLICIES

U.S. FARMS AND CONSERVATION PROGRAMS

AGRICULTURE ISSUES AND POLICIES

Additional books in this series can be found on Nova's website under the Series tab.

Additional E-books in this series can be found on Nova's website under the E-books tab.

ENVIRONMENTAL SCIENCE, ENGINEERING AND TECHNOLOGY

Additional books in this series can be found on Nova's website under the Series tab.

Additional E-books in this series can be found on Nova's website under the E-books tab.

AGRICULTURE ISSUES AND POLICIES

U.S. FARMS AND CONSERVATION PROGRAMS

LUCAS A. SAVARESE
EDITOR

Nova Science Publishers, Inc.
New York

Copyright © 2011 by Nova Science Publishers, Inc.

All rights reserved. No part of this book may be reproduced, stored in a retrieval system or transmitted in any form or by any means: electronic, electrostatic, magnetic, tape, mechanical photocopying, recording or otherwise without the written permission of the Publisher.

For permission to use material from this book please contact us:
Telephone 631-231-7269; Fax 631-231-8175
Web Site: http://www.novapublishers.com

NOTICE TO THE READER

The Publisher has taken reasonable care in the preparation of this book, but makes no expressed or implied warranty of any kind and assumes no responsibility for any errors or omissions. No liability is assumed for incidental or consequential damages in connection with or arising out of information contained in this book. The Publisher shall not be liable for any special, consequential, or exemplary damages resulting, in whole or in part, from the readers' use of, or reliance upon, this material. Any parts of this book based on government reports are so indicated and copyright is claimed for those parts to the extent applicable to compilations of such works.

Independent verification should be sought for any data, advice or recommendations contained in this book. In addition, no responsibility is assumed by the publisher for any injury and/or damage to persons or property arising from any methods, products, instructions, ideas or otherwise contained in this publication.

This publication is designed to provide accurate and authoritative information with regard to the subject matter covered herein. It is sold with the clear understanding that the Publisher is not engaged in rendering legal or any other professional services. If legal or any other expert assistance is required, the services of a competent person should be sought. FROM A DECLARATION OF PARTICIPANTS JOINTLY ADOPTED BY A COMMITTEE OF THE AMERICAN BAR ASSOCIATION AND A COMMITTEE OF PUBLISHERS.

Additional color graphics may be available in the e-book version of this book.

LIBRARY OF CONGRESS CATALOGING-IN-PUBLICATION DATA

U.S. farms and conservation programs / editor: Lucas A. Savarese.
p. cm.
Includes index.
ISBN 978-1-61209-927-9 (hardcover)
1. Agricultural conservation--United States. 2. Farms--United States. I. Savarese, Lucas A.
S604.6.U8 2011
333.76--dc22
2011005855

Published by Nova Science Publishers, Inc. † New York

CONTENTS

PREFACE

Limited-resource and socially disadvantaged farmers make up as much as 40 percent of all U.S. farms. Some federal conservation programs contain provisions that encourage participation by such "targeted" farmers and the 2008 Farm Act furthered these efforts. This new book compares the natural resource characteristics, resource issues and conservation treatment costs on farms operated by targeted farmers with those of other participants in the largest U.S. working-lands and land retirement conservation programs. Some evidence shows that targeted farmers tend to operate more environmentally sensitive land than other farmers, have different conservation priorities, and receive different levels of payments.

Chapter 1- Conservation and commodity programs have many advocates and beneficiaries. Commodity programs support farm families in an effort to ensure abundant supplies of crop commodities; conservation programs encourage stewardship of natural resources and the environment. Can these two aspects of U.S. agricultural policy be joined together into a single, integrated approach to farm support and conservation? Under this hybrid approach, agricultural producers receiving commodity payments would also work to improve their environmental performance (and vice versa)—an appealing quid pro quo. But there is a catch—an integrated program will be effective in achieving both conservation and commodity program goals only if those producers who receive existing commodity payments also face pressing environmental needs.

Chapter 2- Targeted farmers include those with 10 or fewer years of experience, farmers with limited farm sales and income, and farmers belonging to segments of the population that have historically been subject to discrimination, such as African American, American Indian, Alaskan Native, Hispanic, Asian American, or Pacific Island farmers. Targeted farmers make up as much as 40 percent of all U.S. farms, and, although many participate in conservation programs, targeted farmers typically have not participated in government agricultural programs at the same rate as other farmers. To help offset potential barriers to participation, USDA offers targeted farmers more favorable payment and enrollment terms in conservation programs than are available to other farmers.

Chapter 3- Federal working-land and farm and grazing land protection programs have long allocated program funding among States. In 2002, the Regional Equity provision of the Farm Security and Rural Investment Act (the 2002 Farm Act) further emphasized the allocative goal by redirecting some funding to States that historically had received only limited assistance under these programs. The Regional Equity provision mandated that each State receive at least $12 million per year between 2003 and 2007 for enrolling eligible producers in four conservation programs (the 2008 Farm Act increased the threshold to $15 million). This provision may sometimes work at cross purposes to the programs' environmental and economic goals of helping farmers adopt cost-effective conservation practices.

Chapter 4- The Conservation Reserve Program (CRP), enacted in 1985, provides payments to farmers to take highly erodible or environmentally sensitive cropland out of production for 10 years or more. It is the federal government's largest private land retirement program. The program is administered by the Farm Service Agency (FSA) of the U.S. Department of Agriculture (USDA), with technical assistance provided by USDA's Natural Resources Conservation Service. The CRP also has several subprograms, the best-known of which is the Conservation Reserve Enhancement Program (CREP).

In: U.S. Farms and Conservation Programs
Editor: Lucas A. Savarese

ISBN: 978-1-61209-927-9
© 2011 Nova Science Publishers, Inc.

Chapter 1

INTEGRATING COMMODITY AND CONSERVATION PROGRAMS: DESIGN OPTIONS AND OUTCOMES

Roger Claassen, Marcel Aillery and Cynthia Nickerson

ABSTRACT

Can a single program support farm income and encourage producers to adopt environmentally sound farming practices? While simple in concept, attempting to roll the farm income support features of existing commodity programs and conservation payments into a single program raises questions. Exactly how would farm commodity and conservation payments be combined? What difference would it make for environmental gain and farm income support? This chapter approaches the questions in two ways. First, spending patterns in existing commodity and conservation programs are analyzed to determine the extent to which producers who are currently receiving commodity payments also receive conservation payments. Then, a number of hypothetical program scenarios are devised and analyzed to estimate how emphasis on current income support recipients would differ from a combined program that focuses on achieving cost-effective environmental gain. The results show that policymakers face significant tradeoffs between environmental (conservation) objectives and farm income support objectives in designing a program that provides both income support and environmental gain.

Keywords: conservation, commodity programs, income support

ACKNOWLEDGMENTS

Shawn Bucholtz, Nathaniel Higgins, and J. Michelle Michalek also contributed to this paper. We thank Sandra Batie, Cathy Kling, Phil Spinelli, John Stierna, Marca Weinberg, and Ed Young for helpful comments on earlier drafts. Thanks also to Priscilla Smith for editorial assistance and Wynnice Pointer-Napper for layout and design.

SUMMARY

Conservation and commodity programs have many advocates and beneficiaries. Commodity programs support farm families in an effort to ensure abundant supplies of crop commodities; conservation programs encourage stewardship of natural resources and the environment. Can these two aspects of U.S. agricultural policy be joined together into a single, integrated approach to farm support and conservation? Under this hybrid approach, agricultural producers receiving commodity payments would also work to improve their environmental performance (and vice versa)—an appealing quid pro quo. But there is a catch—an integrated program will be effective in achieving both conservation and commodity program goals only if those producers who receive existing commodity payments also face pressing environmental needs.

What is the Issue?

Policymakers may need to compromise commodity program objectives, conservation objectives, or both, in merging conservation and farm commodity payments into an integrated "green payments" program.

This chapter:

- examines the extent to which participation in existing conservation and commodity programs overlap
- devises a set of hypothetical scenarios covering a wide range of possible green payment program designs
- analyzes likely producer reactions and the resulting environmental and income support outcomes for each of these scenarios.

What Did the Study Find?

Policymakers may face significant tradeoffs if they attempt to combine farm commodity and conservation payments. Commodity payments are intended to support farm families while ensuring abundant supplies of crop commodities at competitive market prices. These goals are quite distinct in scope and emphasis from those of conservation programs, which are designed to promote environmentally sound farming practices. Many farms that receive existing conservation payments or offer opportunities for cost-effective conservation do not receive payments from existing commodity programs.

In 2004, only a small proportion of U.S. farms (6 percent) received payments from both commodity and conservation programs, partly because conservation programs have been small relative to commodity programs. Because conservation program budgets and payments are increasing, however, the overlap between commodity and conservation payments is likely to increase. Nonetheless, only 43 percent of conservation program payments in 2004 went to farms that also received commodity program payments. This suggests that a significant share of new conservation payments could go to farms that currently do not receive commodity payments.

Conservation and farm commodity payments could be combined in many ways. In devising hypothetical scenarios, we consider variations on two general approaches—environmental compliance and environmental performance. The approaches are selected to help identify and characterize possible tradeoffs between conservation and the farm income support features of existing commodity programs in designing an integrated program, as follows:

Environmental Compliance. Policymakers could start with existing commodity programs and make them "greener" by adding environmental compliance requirements. While the new requirements could result in greater conservation effort by commodity payment recipients, producers' additional conservation costs would cut into income support. Moreover, current farm commodity payments reach only about 25 percent of U.S. farms, although those farms control about 80 percent of cropland and 50 percent of all agricultural land. The other 75 percent of farms and ranches, including many with pressing environmental needs, are not eligible for payments because they do not produce program crops.

Environmental Performance. On the other hand, policymakers could start from a conservation program perspective, devising a set of conservation payments that could exceed producers' conservation costs and, therefore, support farm income. One way to do that is to offer payments that are commensurate with environmental performance rather than cost. To the extent that payments exceed cost, producers could make a "profit" on producing environmental gains. Because program eligibility would not be confined to farms that receive commodity payments, however, income support and conservation effort would be spread more broadly across the farm sector than for the environmental compliance scenarios. If policymakers decide to offer an integrated green payment program in lieu of existing commodity programs (rather than in addition to these programs), current recipients of commodity payments could realize a loss in net income support.

Empirical analysis shows both similarities—and significant differences— across the hypothetical scenarios. In general, we estimate that:

- **Both environmental compliance and environmental performance scenarios deliver both environmental gain and income support.** While neither approach assumes any specific funding levels for income support or conservation, both can produce substantial income support and environmental gain. Depending on the specific scenario, conservation expenditures account for as much as 50 percent of total payments to producers. The balance of the total payment (total payment less net conservation expenditures) is income support.
- **Environmental gain depends critically on program design.** While both environmental compliance and environmental performance scenarios leverage environmental gain, the environmental performance scenarios realize gains at a lower cost per unit of environmental gain. In other words, environmental performance scenarios are more cost-effective than the environmental compliance scenarios in producing environmental gain. More cost-effective environmental gain means that a given budget can produce more environmental gain, more income support, or both.

- **Policymakers may face a difficult tradeoff between environmental gain and the *distribution* of income support.** Cost-effective environmental gains are achieved largely by encouraging the enrollment of producers who can deliver large environmental gains per dollar of cost. These producers, however, are not necessarily those historically receiving commodity program payments. If policymakers want to continue supporting recipients of existing commodity program payments, they are likely to face a difficult tradeoff between environmental gain and income support.

How Was the Study Conducted?

Analysis of existing commodity and conservation payments is based on Agricultural Resources Management Survey (ARMS) data for 2004. To analyze the hypothetical program scenarios, the authors developed a model based on the ARMS farm business and household survey for 2002. Data from 2002 are used because it is the most recent year for which grazing land acreage is provided. Additional data on conservation treatment needs (e.g., whether soil erosion, nutrient runoff, etc., are problems) and the potential for environmental gains were also used. Data sources include the National Resources Inventory and the Workload Assessment data, both maintained by USDA's Natural Resources Conservation Service (NRCS). To quan-tify environmental gain, we used an environmental index, similar to the Environmental Benefits Index used in the U.S. Department of Agriculture's Conservation Reserve Program. Producers were assumed to participate if payments exceed the minimum payment they would accept for undertaking a given treatment or set of treatments. The payment needed to make farmers willing to adopt specific treatments was estimated from NRCS Environmental Quality Incentives Program data.

Supporting Farm Income and the Environment: Can a Single Program Do Both?

Can a single program provide income support similar to existing commodity programs and improve the environmental performance of U.S. farms? An integrated "green payment" program would attempt to do both—combining key elements of existing farm commodity and agricultural conservation programs. Although the idea is not exactly new—some existing programs do support income and encourage better environmental performance—most programs focus on commodity support or conservation and have a secondary or limited effect on the other. Existing farm commodity programs, for example, are intended to support farm families historically involved in the production of major field crops, but also link payments to environmental compliance requirements. To maintain eligibility for commodity program payments, producers must apply approved conservation systems on highly erodible cropland and refrain from draining wetlands for crop produc-tion. Perhaps the best existing example of a conservation program that also provides income support is the Conservation Security Program (CSP), administered by USDA's Natural Resources Conservation Service. Producers can receive "stewardship" payments based on past conservation efforts rather than current conservation costs. While only about 15 percent of program funds are devoted to stewardship payments, these payments could enhance farm income because they are not tied to the cost of adopting or maintaining conservation practices.

A more complete merger of the income support features of existing commodity programs and conservation payments—an integrated "green payment" program—could be pursued for a variety of reasons. As the importance of conservation programs in overall U.S. farm policy rises, for instance, green payments could be seen as a way to harness commodity program payments for environmental gain. One way to do that would be to raise the bar on environmental compliance. Additional compliance requirements could include soil conservation (on cropland that is not considered highly erodible), nutrient management, or pest management. With this approach, a green payment program would continue to focus on traditional recipients of commodity program payments.

Environmental performance or stewardship could also be seen as a primary basis for farm program payments. At present, about 25 percent of U.S. producers receive commodity program payments; these producers account for more than 80 percent of cropland and 65 percent of crop production (by value). If producer payments were based, instead, on some measure of envi-ronmental performance, income support could be available to a broader range of producers and could leverage a broader range of environmental gains when compared to an expansion of compliance requirements for traditional commodity programs.

Can green payments be an effective mechanism for delivering both income support and environmental gain? Because green payments, as we use the term in this chapter, would join two important aspects of U.S. agricultural policy, it is tempting to view them as a "win-win" proposition, perhaps increasing support of agricultural policy or even saving money. Inevitably, however, tradeoffs will arise. The portion of a green payment that producers use to pay conservation costs (e.g., adopting and maintaining conservation practices) will not support farm income. Income will be supported only to the extent that payments exceed conservation costs. The proportion of payments that would be used for conservation as opposed to income support is an important issue in green payment design.

THE MANY SHADES OF GREEN

The term "green payment" has been used in a number of different contexts to mean different things. As we use the term, a green payment is based on a (relatively) co-equal consideration of both income support and conservation (environmental) objectives. Many individuals and organizations concerned about farm policy in general and agricultural conservation policy in particular also use the term in this way. Others, however, view green payments as referring to any conservation or environmental payment, regardless of its relationship to income objectives. We refer to these as conservation payments. Still others view green payments as agricultural payments of any type that meet the criteria for inclusion in the World Trade Organization (WTO) "green box." In this context, being green isn't about environmental performance. Policies end up in the WTO green box if they have little or no impact on commodity prices or trade. These programs are given the *green light* to go forward under WTO rules, but are not considered within our usage unless they provide both income support and conservation assistance to farms.

An equally important question is whether farms that receive income support under existing commodity programs can also make cost-effective contributions to improving environmental quality. If farms historically eligible for commodity program payments would not otherwise be targeted for conservation payments, prioritizing farms for green payments could involve compromising income support objectives, conservation objectives, or both. While most farms could address one or more environmental issues or "resource concerns," the potential for environmental gain can vary across farms depending on crops, production practices, climate, location, and other factors that ultimately determine the effect of agricultural production on the environment.

Green Payments and the WTO

Would green payments be eligible for the WTO "green box?" The "green" in "green box" is an analogy to traffic lights: Green means go. In WTO parlance, a green box program does not distort trade or causes only minimal distortions and may be implemented freely by member nations. Some have argued that green payments would be a WTO green-box-compliant way to support farm income. An analysis of conservation policy options, developed by the Secretary of Agriculture in 2006, raises important questions about green payments and the green box (USDA, 2006).

WTO rules do allow "green box" options for both income support and conservation programs. Income support payments that are "decoupled" from (not dependent on) *current* production, prices, and input use can qualify for the green box. Working-land conservation programs can also qualify for green box status if payments are for a clearly defined conservation or environmental purpose and do not exceed the extra cost or lost income directly related to conservation activities. A number of U.S. conservation programs, including the Environmental Quality Incentives Programs, have been notified (reported) to WTO under these provisions. Eligible conservation programs generally provide partial reimbursement (cost-sharing) or incentive payments designed to be equal to or less than producer costs.

Would green payments satisfy green box requirements? If green payments were to be reported as conservation payments, they could qualify if producers are compensated only for their costs or income forgone. But that would preclude income support—payments would simply offset producer costs. If green payments were to be reported as decoupled income support, on the other hand, payments could not be dependent on current input uses. Changes in input uses, however, are often the means by which conservation payments leverage environmental gains, so removing requirements for such changes may diminish their environmental effectiveness. In short, a green payment program that provided both income support and environmental gains would not necessarily qualify as a green box program under WTO rules.

If existing commodity program payments and opportunities for cost- effective conservation do not occur largely on the same farms, funds devoted to conservation on farms that receive commodity program payments are likely to produce some environmental gain, but would likely produce more environmental gain per dollar spent if applied elsewhere. Green payments could support income on farms where the potential for cost-effective environmental gain is high, but these farms may or may not be the farms that have traditionally received farm income support through commodity program payments.

Finally, the treatment of producers who have already achieved a relatively high level of environmental performance is an issue. Some have argued that excluding these producers would be inequitable and could create incentives to defer conservation action. If producers who have adopted conservation practices without government payments are excluded from conservation programs (for conservation work they have already completed), they may be discouraged from going forward with conservation work in the absence of payments. In the long run, moreover, payments that exclusively subsidize change in environmental performance will eventually result in a withdrawal of support from farms that have made a great deal of environmental improvement. If policymakers want to continue supporting these producers, these already established "good actors" would have to be eligible for ongoing payments.

EXISTING CONSERVATION AND INCOME SUPPORT PROGRAMS: DIFFERENT PURPOSES, DIFFERENT PAYMENTS, DIFFERENT PRODUCERS

Existing conservation and farm commodity programs have different purposes, which lead to fundamental differences in how the two types of programs are structured and administered. Commodity-based income support is intended to support farm families historically involved in the production of targeted crops by enhancing the incomes of eligible producers, primarily the producers of major field crops—corn, wheat, soybeans, cotton, and rice. Historically, producers with larger production received larger payments. Since 1996, some (but not all) commodity program payments have been based on historical crop acres and yields rather than current acres and yields. The change was designed to reduce the effect of commodity payments on production decisions and avoid stimulating overproduction. Even so, producers who farm highly productive land (with a history of high yields) that is eligible for commodity payments (by virtue of a history of program crop production) will tend to reap the largest payments. For more details, see www.ers.usda.gov/Briefing/ FarmPolicy/DirectPayment s.htm/.

Conservation payments, on the other hand, are designed to prompt change in land use or production practices to have a beneficial environmental effect. Conservation payments are available to a wider range of producers—nearly all crop and livestock producers are eligible for at least one conservation program. While conservation programs seek to change production practices, the level of production may or may not be affected. Land retirement is likely to affect production, although how much depends on the quality of the land retired and the extent to which other land is converted to crop production (sometimes referred to as "slippage"). But, many conservation practices have little or no impact on production levels. Producers who install terraces to reduce soil erosion, for example, are likely to see little change in production, at least in the short term. Most conservation payments are limited to the amount necessary to prompt adoption of new practices, perhaps covering only a portion of the producer's cost through cost-sharing. Some programs use competitive bidding among producers to stretch program budgets.

About 40 percent of U.S. farms, representing 60 percent of all agricultural production (by value), receive some type of government payment. Of that 40 percent, 15 percent—about 6 percent of all U.S. farms—received both commodity (income support) and conservation payments in 2004 (figure 1). Since 2002, conservation program funding has increased considerably, particularly through working land programs like the NRCS Environmental Quality Incentives Program (EQIP). Because actual payments to farmers often come several years after EQIP enrollment (as specified conservation work is completed), it is likely that the number of farms receiving both commodity and conservation payments will also increase in coming years. In 2004, however, less than half of conservation payments (43 percent) went to farms that also received commodity payments, so a large share of additional conservation payments could also flow to farms that do not receive commodity payments, including many specialty crop and livestock farms.

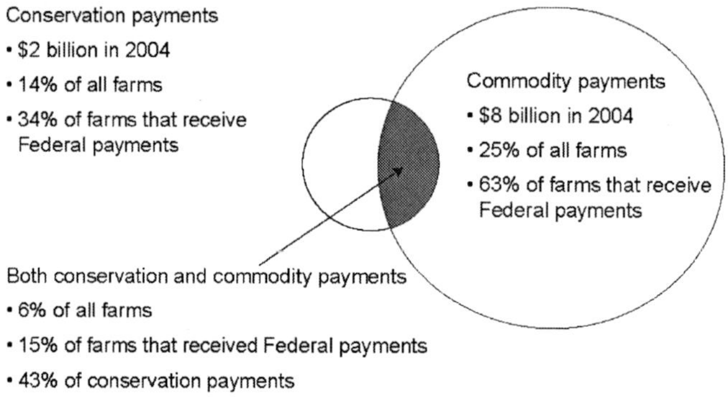

Source: USDA, Agricultural Resource Management Survey data, 2004.

Figure 1. Overlap between income and conservation payment recipients is small.

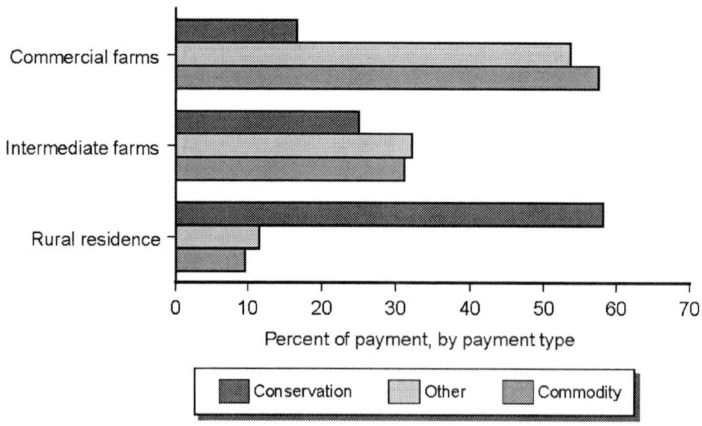

Note: Other payments are largely ad hoc agricultural disaster payments.

[1] Collapsed ERS farm typology divides farms into three groups: (1) commercial farms are large with sales above $250,000; (2) intermediate farms have sales below $250,000 and the operator reports farming as his or her major occupation; and (3) rural residence farms have gross sales below $250,000 where farming is considered a secondary occupation or activity.

Source: USDA, Agricultural Resource Management Survey data, 2004.

Figure 2. Distribution of Federal agricultural payments by collapsed ERS farm typology[1].

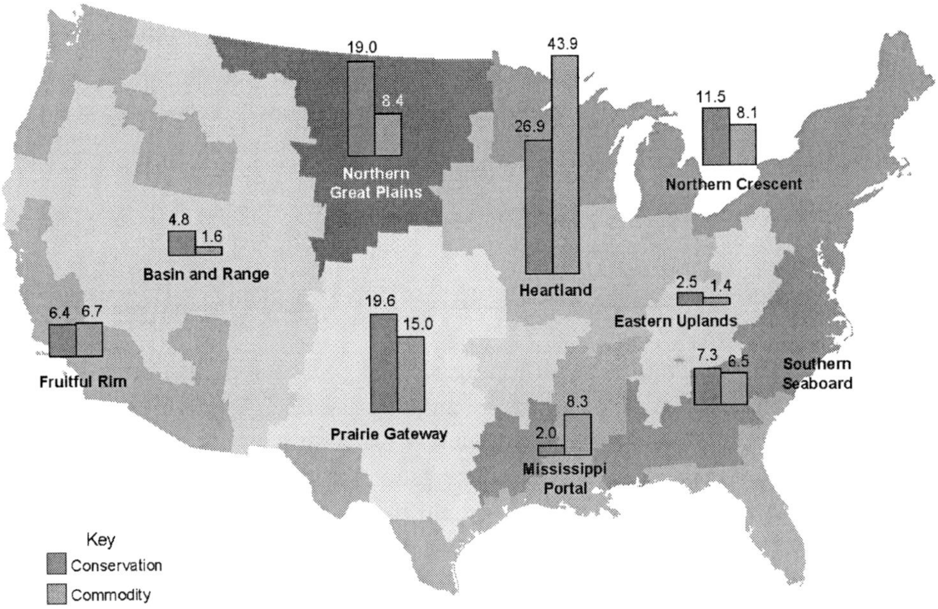

[1] ERS farm resource regions are explained in *Farm Resource Regions*, AIB760. http://www.ers.usda.gov/ Publications/AIB760/.

Source: USDA, Agricultural Resource Management Survey data, 2004.

Figure 3. Regional shares of commodity and conservation payments (percent of national total)[1].

Differences in the distribution of commodity and conservation payments across farm types and regions in the United States are striking. Most commodity program payments go to large, commercial farms, while most conservation payments go to rural residence farms[1] (figure 2). Commodity payments are concentrated in areas where production of program crop commodities is prevalent—the Corn Belt, Northern Plains, and the Mississippi Delta (figure 3). Conservation payments tend to be concentrated in some areas of the Northern Great Plains, Prairie Gateway, Northern Crescent, and Basin and Range regions (figure 3).

To some extent, the minimal overlap between conservation and commodity payments means that environmental and income support priorities are leading these programs to focus on different producers in different regions. The existing distribution of conservation payments largely reflects a historical reliance on land retirement to attain conservation goals. Rural residence farms are more likely than other farms to retire land from crop production through government programs such as the Conservation Reserve Program (CRP), but less likely to receive farm income support payments. It is possible that these farms are more likely to be located on CRP-eligible land, although existing data are not sufficient to test this possibility. Another possible—but untested—explanation is that these farms are more willing than other farms to give up crop production (e.g., some producers may have decided to retire or seek other employment given the opportunity to enroll land in CRP). In any case, high levels of CRP participation are responsible for the fact that a large share of conservation payments flow to rural-residence farms.

In recent years, conservation program funding has risen rapidly, largely through increased funding for working land programs, primarily EQIP. As the proportion of conservation dollars spent through EQIP rises, a larger proportion of conservation dollars may also go to commercial and intermediate farms, although existing data are not sufficient to draw a strong conclusion on this point. Even if a larger proportion of EQIP funding does go to commercial and intermediate farms, however, 60 percent of funding for EQIP must, by statute, address livestock-related issues. Livestock farms are less likely than crop farms to receive payments through existing commodity programs. It is not clear how EQIP money would be distributed without the 60-percent requirement.

Green Payment Program Design: A Matter of Perspective

While analysis of existing commodity program and conservation payments suggests the potential for tradeoffs between income support and conservation objectives, some features of existing conservation programs make it difficult to reach strong conclusions based on this experience alone. For example, current conservation program funding emphasizes land retirement, while a green payments program would likely focus on encouraging conservation practices on working land. To gain additional insight, we analyze a number of hypothetical green payment program designs.

"Program design" refers to the details of a program: who is eligible, what action or activity producers could be paid for (e.g., conservation treatments), and how much they would receive for specific actions. While all green payment program designs would seek to support farm income and improve environmental performance, one could approach design decisions from a number of perspectives. To identify potential tradeoffs between income support to current recipients and environmental gain, we analyze some "polar" cases—program designs that originate from decidedly different perspectives. On the one hand, policymakers could start from a primarily environmental point of view, establishing a set of environmental payments that are large enough to leverage environmental gain and provide income support. These are *Environmental Performance* scenarios. On the other hand, policymakers could focus on the recipients of current commodity program payments, making these payments "greener" through the addition of environmental requirements, similar to (but going beyond) existing conservation compliance, sodbuster, and swampbuster requirements. (Sodbuster and swampbuster are designed to discourage producers from bringing additional highly erodible land and wetland, respectively, into crop production.) These are *Environmental Compliance* scenarios.

Some key details of program design are common to all four outlined scenarios. First, all scenarios assume that farmers are offered 5-year contracts and "program payments" are generally represented as the sum of all payments over the 5-year period. Second, because income support is a primary objective, we assume that a green payment program would be run as an entitlement, in keeping with existing farm commodity programs. Under an entitlement (like existing commodity programs), the Government is obliged to enroll producers who apply for the program and qualify for benefits. Program spending depends, in part, on the level of producer participation. In contrast, existing conservation programs are limited by an annual budget or acreage cap that limits the number of producers and acres that can be enrolled. In our green payment scenarios, the Government would establish rules governing eligibility and the calculation of payments, but the exact level of program payments for any specific scenario is determined by the number of farms that participate, the number of acres they choose to enroll, and the conservation treatments they apply.

In the *Environmental Performance* scenarios, farmers and ranchers are offered the opportunity to (voluntarily) produce environmental "goods" for a "price" established by the Government. Environmental goods could include clean water or wildlife habitat that farmers and ranchers produce by applying conservation treatments. For example, they could help produce clean water by controlling runoff of sediment, nutrients, and/or pesticides from agricultural land. See table 1 for a full listing of the resource concerns that could be addressed in our green payment scenarios and the conservation treatments that might be used to address them.

In keeping with existing conservation programs, producers are allowed to determine (within guidelines) which land and conservation treatments they will offer for green payment program participation. All cropland and grazing land could be eligible for program enrollment, so long as the proposed conservation treatment would address a specific resource concern present on or associated with the tract being offered. The payment a producer could receive for taking these actions would be roughly proportional to his or her probable contribution to the production of environmental goods. Our analysis uses an environmental index, similar to the Environmental Benefits Index (EBI) used in the CRP, to quantify a producer's environmental performance and estimate the gain in environmental performance from any given conservation treatment (see box, "The Environmental Index," and appendix 1 for full details). Specific scenarios include:

Table 1. Linking resource concerns, land use, physical effects, and treatments.

Resource concern	Physical effect	Land use	Treatment
Air quality	Wind erosion	Cropland[1]	Wind erosion control
Surface-water quality	Water erosion Nitrogen runoff Phosphorus runoff Nutrient runoff and riparian erosion Pesticide runoff	Cropland Cropland Cropland Grazing land Cropland	Water erosion control Nutrient management Nutrient management Nutrient management and riparian erosion control Pest management
Groundwater quality	Nitrogen leaching Pesticide leaching	Cropland Cropland	Nutrient management Pest management
Soil productivity	Wind erosion Water erosion	Cropland Cropland	Wind erosion control Water erosion control
Grazing land health	Grazing land health	Grazing land	Grazing land health
Wildlife	Wildlife habitat loss or degradation	Cropland Grazing land	Habitat restoration or enhancement Habitat restoration or enhancement

[1] Irrigated and nonirrigated cropland are combined here but are treated separately in our analysis.

Source: USDA, Economic Research Service.

The Environmental Index

To base payments on environmental performance, some method of measuring performance is needed. Environmental indexes are used widely in conservation programs to gauge the potential environmental gain from the application of conservation treatments. Indexes combine data on a number of environmental dimensions into a single number. In several USDA programs, including the Conservation Reserve Program (CRP) and the Environmental Quality Incentives Program (EQIP), program managers use indexes to rank contracts for program enrollment. For a limited number of resource concerns, the Conservation Security Program also offers payments that vary according to improvement in the value of indexes believed to reflect environmental performance. In our model of green payments, we develop an overall index of environmental performance and use it to specify environmental performance-based payments.

Our index is similar to the Environmental Benefits Index (EBI) used to rank contract offers in the CRP general signup:

- Roughly one-third of points are for soil erosion on cropland. Points are given for potential of erosion control to reduce dust (improve air quality), preserve soil productivity, and reduce sediment loads to water.
- Another third are for other water quality-related treatments, including nutrient management and pest management on cropland, nutrient management and riparian erosion on grazing land, and grazing land health.
- Remaining points are for wildlife habitat enhancement (on cropland and grazing land).

The share of possible points assigned to a specific farm depends on the intensity of the physical effects (e.g., soil erosion or nutrient runoff) and the potential damage to soil, water quality, or other resources. For example, if soil erosion due to water (tons per acre per year) is estimated to be high on a field located in an area where water-quality damage per ton of soil erosion is also estimated to be high, a large share of potential points would be assigned for the index subcomponent that accounts for sediment damage to water quality. A complete description of the index can be found in appendix 1.

- ***Improved Performance.*** Payments would be based on expected environmental gain, as measured by our environmental index. Producers would receive payments based on the application of *additional* conservation treatments that yield a *gain* in environmental performance. The payment made to a producer would equal the change in the producer's environmental index score (no matter what his or her starting point), multiplied by a payment rate per index point determined by the Government and announced to producers as part of the program signup notice.
- ***Good Performance.*** Payments would be based on environmental *performance*. Once producers reach a predetermined level of environmental performance—which we refer to as an environmental hurdle—they are eligible for payments. Producers do not necessarily need to apply new conservation treatments—they can qualify for payments even if environmental performance was achieved before the establishment of the green payment program. The hurdle rate is set by region so that about half of all agricultural land in each region qualifies for payments without additional

conservation effort.[2] Producers who have already surpassed the hurdle can also increase payments by further improving environmental performance (undertaking additional conservation treatments). For an individual producer, payment would be equal to the difference between his or her index score and the environmental hurdle, multiplied by the payment rate per index point. That is, the payment equals the payment rate times the difference between the index score and the hurdle rate, if the index score is greater than the hurdle rate. If the index score is less than the hurdle rate, the payment is zero.

Table 2. Summary of scenarios for green payment analysis.

Scenario type	Scenario	Eligible farms/land	Payment "trigger" (action/condition)	Payments
Environmental Performance	1. Improved Performance	All cropland and grazing land; farms that include either cropland or grazing land	Any additional treatment appropriate to the farm	Based on environmental gain, as measured by the change in environmental index score
	2. Good Performance		Environmental performance exceeds a predetermined "hurdle" rate	Based on environmental performance, as measured by environmental index, relative to the hurdle rate
Environmental Compliance	3. Extended Compliance	Cropland on farms that receive income support	Meet existing compliance requirements and control soil erosion on all land, manage nutrients; pests	Similar to existing direct payments; producers must meet all conser-vation treatment requirements to maintain eligibility
	4. Modified Compliance		Meet existing compliance requirements and control soil erosion on all land, manage nutrients; pests	Producers can opt out of someconservation treatment require-ments for a reduction in payments, commensurate with reduction in environmental performance, as measured by environmental index

Source: USDA, Economic Research Service.

In the *Environmental Compliance* scenarios, existing farm commodity payments are used to leverage additional conservation effort and improve environmental performance. Existing compliance requirements for wetland conservation and soil conservation on highly erodible cropland (HEL), in force since 1985, would remain unchanged, while new compliance requirements would be added. New compliance requirements would include reducing soil erosion to the soil loss tolerance ("T" level) on non-HEL cropland, as well as nutrient management and pest management on cropland.[3] Specific scenarios include:

- ***Extended Compliance.*** Continued eligibility for commodity program payments would be contingent on addressing all existing and new compliance requirements, regardless of cost. Producers who do not meet all applicable requirements would become ineligible for commodity program payments, and the income support they provide, on all the land they farm.
- ***Modified Compliance.*** Producers could opt out of some of the new requirements if they accept a reduced payment. Producers who address no additional requirements would still receive 20 percent of the maximum payments they would be eligible for. Producers who address all additional requirements would receive 100 percent of their potential payment. For producers who opt out of some, but not all additional requirements, payment reduction would be commensurate with the environmental gain forgone, as measured by our environmental index.

Producer Participation: Doing the Math

Farm income and environmental outcomes depend largely on whether (and how) farmers and ranchers choose to participate in a green payment program. Imagine a producer sitting down to "pencil out" his or her green payment program options. He or she may have a number of tracts of land that could qualify for payments if one or more resource concerns were addressed. Addressing a resource concern would entail the application of an appropriate conservation practice or a set of practices (sometimes referred to as a conservation "treatment"; see table 1). The existence of a water-quality concern, for example, may lead to treatment for soil erosion (to reduce sediment flows), nutrient runoff, or pesticide runoff through conservation tillage (and other soil erosion control practices), nutrient management, or pest management, respectively.

The producer's participation decision boils down to a single question: Am I willing to accept the payment offered in exchange for undertaking the prescribed conservation treatment(s)? Producers may consider a range of factors in deciding whether a given payment is large enough:

- out-of-pocket costs
- changes in production (e.g., change in crop yields)
- difficulty of managing and maintaining required conservation practices
- changes in production risk (e.g., an increase or decrease in the probability of low yields).

All of these factors come together in the producer's willingness to accept (WTA), defined as the minimum payment he or she would be willing to accept in exchange for taking a specified action. Farmers and ranchers will participate any time the prospective payment exceeds their WTA for the conservation treatment in question. If a producer is willing to accept $4 per cropland acre for nutrient management, for example, he would agree to undertake nutrient management if the per-acre payment for nutrient management is $4 or higher. For ease of exposition, we also refer to WTA as "economic cost" because all factors outlined above are real (or economic) costs to the producer, even if they are not out-of-pocket costs. From a Government perspective, we also refer to economic costs as net conservation expenditures because it is the minimum the Government must pay to leverage a specific conservation action on a specific farm. Payments in excess of economic cost (WTA) are income support.

We simulate the process of "penciling out" green payment options for each one of a series of model farms, based on the 2002 Agricultural Resources Management Survey (ARMS):

- For each green payment program scenario defined above, we simulate a set of green payment participation options for each farm.[4] The options are based on the requirements of the scenario, e.g., the types of land and conservation treatments that trigger payment, and a farm-specific estimate of the number of acres that could be treated (see appendix 2 for details).
- The level of payment the producer would be willing to accept for undertaking any specific option is estimated using contract data from the Environmental Quality Incentives Program and techniques detailed in appendix 3.
- In general, we assume that the producer selects the option that yields the largest net return over economic costs or net income support. In other words, we define net return and net income support as the total payment to the producer less the producer's economic cost for undertaking prescribed conservation treatments (see appendix 4 for a detailed discussion of producer decision rules). We also assume that producers will participate only when net return exceeds $200 (total) to offset transaction costs, e.g., the cost of application and related expenses.

Because the scenarios are analyzed as entitlements, total payments to producers are a function of producer response to the payment rates offered, rather than a program budget. To compare our scenarios across a wide range of program sizes (i.e., total producer payments), we vary these payment rates. For the *Improved Performance* and *Good Performance* scenarios, program payments are varied implicitly by varying the payment rate per environmental point. As the payment rate rises, more producers participate, and those who would have participated at a lower payment rate undertake additional conservation treatments. Because the scenarios are different, however, the payment rate corresponding to a given level of program payments varies across scenarios.

For the *Extended Compliance* and *Modified Compliance* scenarios, program payments are also varied implicitly by varying the farm-specific level of direct payments (as reported for the ARMS farms that are the basis for the analysis) using a scale factor. For example, if the scale factor is 1.2, each farm would receive (at most) 120 percent of the direct payment the farm received in 2002. Given the economic cost of complying with environmental compliance requirements, participation in the environmental compliance scenarios increases as the scale factor increases. For *Modified Compliance*, moreover, producers also undertake additional conservation treatments— treatments they may otherwise have opted out of—as the scale factor rises.

Green Payment Program Design Tradeoffs: Do Income Support and Environmental Gain Go Well Together?

While there are many ways to combine conservation and commodity payments, the real question is whether they are a good match. The answer depends largely on policymakers' conservation and income support objectives. Using the distribution of payments under existing commodity programs as a benchmark, however, we can analyze some key questions that policy- makers would almost surely face in designing a green payment program:

- What proportion of green payments ends up as income support?
- How much environmental gain is obtained given the level of net conservation expenditure?
- How does conservation cost-effectiveness affect the *distribution* of income support?

What proportion of green payments ends up as income support? At some level, there is always a tradeoff between income support and the environment. In a green payment context, policymakers implicitly relinquish control over the allocation of funds between income support and environmental gain in order to merge these two program objectives. The portion of producer payments that offsets producer economic costs cannot support farm income. Once the economic cost of required conservation actions are covered, the remainder of the green payment is net farm income support.

Both *environmental compliance* and *environmental performance* scenarios deliver environmental gain and income support. The portion of total producer payments that covers the economic costs of taking specified conservation actions varies from 10 percent to as high as 50 percent, depending on the program scenario and the overall level of program payments. The balance of the payments—net income support —ranges from 50 to 90 percent of overall program payments (figure 4). The ultimate effect of this split between conservation and income support expenditures on overall income support and conservation effort also depends on whether a green payments program is in addition to or instead of existing programs.

At an aggregate level, the tradeoff between net income support and net conservation expenditure is modest. When program payments are relatively low (less than $5 billion over 5 years), the *Good Performance* and *Modified Compliance* scenarios yield the highest overall levels of income support. That's because producers can participate without additional environmental effort. Low payments result from low payment rates, prompting many producers to base their participation on existing conservation efforts rather than new conservation action, if possible. So, payments are largely devoted to income support. In contrast, income support is relatively low for the *Improved Performance* and *Extended Compliance* scenarios, where producers must take additional action to receive payments.

When payments are larger—$20 billion or more over 5 years—the compliance scenarios yield the largest income support, although the difference between the environmental compliance and environmental performance scenarios is not dramatic. The proportion of payments going for income support in the environmental compliance scenarios rises quickly as payments become large, reflecting the fact that limiting eligibility (to recipients of existing commodity program payments) also limits opportunities for conservation treatment. In other words, as the scale factor in the environmental compliance payments rises, there is a shrinking pool of eligible acres that still need conservation treatment.

How much environmental gain is obtained given the level of net conservation expenditure? That is, how cost-effective is each scenario in terms of environmental gain per dollar of conservation spending? A specific green payment program design is cost-effective *relative to* another design if it produces more environmental gain for a given level of net conservation expenditure (*not* total program payments, which also include an income support component). Because the overall level of income support lies in a relatively narrow range across green payment scenarios, particularly when total payments are $5 billion or larger, net conservation expenditures also lie within a relatively narrow range. Given similar levels of net conservation expenditure, differences in environmental gain depend largely on the cost effectiveness of conservation expenditures.

Environmental gain depends critically on program design. Figure 5 shows net income support and environmental gain (in terms of environmental points) for our four hypothetical green payment programs at three different levels of overall program payments. The oval in the lower left-hand side of the plot area shows net income support and environmental gain for each scenario when program payments are $5 billion (total over 5 years). Other ovals correspond to total payments of $10 billion and $15 billion.

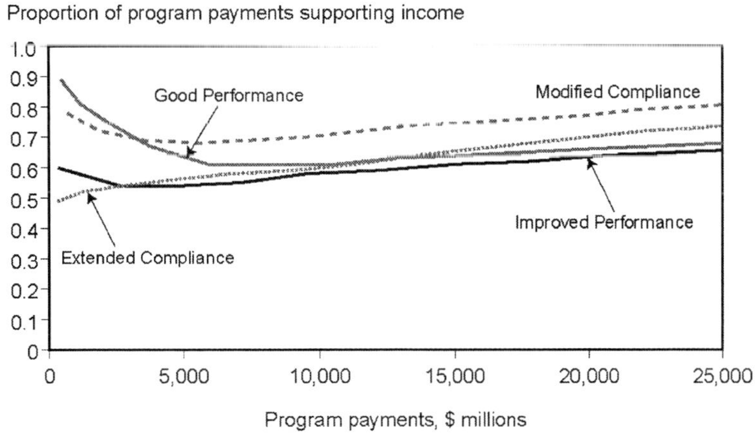

Source: USDA, Economic Research Service.

Figure 4. Net income support as a proportion of program payments.

As a point of comparison, we also graph a fifth scenario in which all funds are channeled into environmental gain. We refer to this scenario as *Environmentally Efficient*. The points representing various levels of program payments are located on the vertical (environmental points) axis in figure 5. This scenario is identical to *Improved Performance* except that payments are only large enough to cover producer economic costs (WTA). In other words, the entire payment goes to leverage conservation; income support is zero. In theory, competitive bidding can yield payments just large enough to cover producer conservation cost, although it would be very difficult to design and implement an auction in which all producers submitted bids equal to their economic costs. Nonetheless, this scenario gives an upper bound for the potential of a purely environmental program.

The downward sloping curves in figure 5 represent all combinations of income support and environmental gain that could be achieved by separate programs focusing, respectively, on income support and conservation, given $5 billion, $10 billion, and $15 billion in total payments (the sum of conservation expenditure and net income support). We refer to these curves as cost-effectiveness frontiers. Suppose fixed budgets of $5 billion, $10 billion, and $15 billion are available for allocation between the *Environmentally Efficient* scenario and an income support program similar to the existing direct payment program. By varying the allocation of funds across the two programs and using our model to estimate the maximum possible environmental gain, we define all feasible combinations of income support and environmental gain, given the available budget. Combinations of environmental gain and income support that are on or inside (below and the left of) the cost- effectiveness frontiers can be achieved given $5 billion, $10 billion, and $15 billion in program payments.

These cost-effectiveness frontiers show the underlying tradeoff between environmental gain and income support when separate, cost-effective programs are used to leverage environmental gain and provide income support. Following the $10-billion frontier beginning at the horizontal axis in figure 5, shifting funds from income support to the *Environmentally Efficient* scenario would increase environmental gain, rapidly at first (the frontier is almost vertical) indicating that some environmental gain can be achieved at very low cost. As more money is shifted from income support to environmental gain (moving toward the upper left) additional environmental gain declines as indicated by the decrease in the slope of the cost-effectiveness frontier. The increase in the cost of additional environmental gain is driven by the fact that payment incentives encourage producers to undertake the least expensive (most cost-effective) gains first.

Figure 6 is similar to figure 5, but shows treated acreage rather environmental points. Figure 6 shows that treated acreage is not necessarily a good indicator of environmental gain. While some scenarios (other than *Improved Performance*) can treat as many or even more acres than could be treated with the *Environmentally Efficient* scenario, different acres would be treated—acres that would produce less environmental gain as measured by our environmental index. Environmental targeting generally produces this type of result—policymakers choose to pass up acres that could be treated cheaply in favor of treating acres that produce large environmental gain relative to treatment cost.

Returning to figure 5, note that only the *Improved Performance* and *Environmentally Efficient* scenarios are on the cost-effectiveness frontiers.

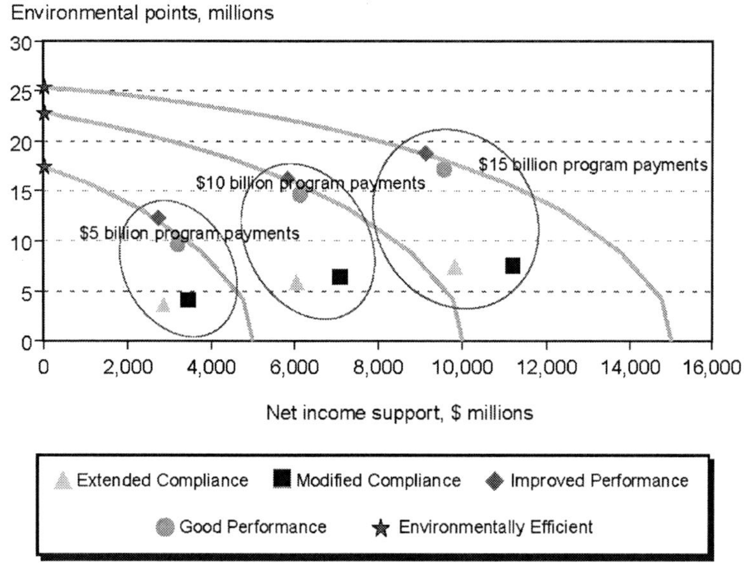

Source: USDA, Economic Research Service.

Figure 5. Efficiency matters: Trading environmental gain and income support in green payment program design.

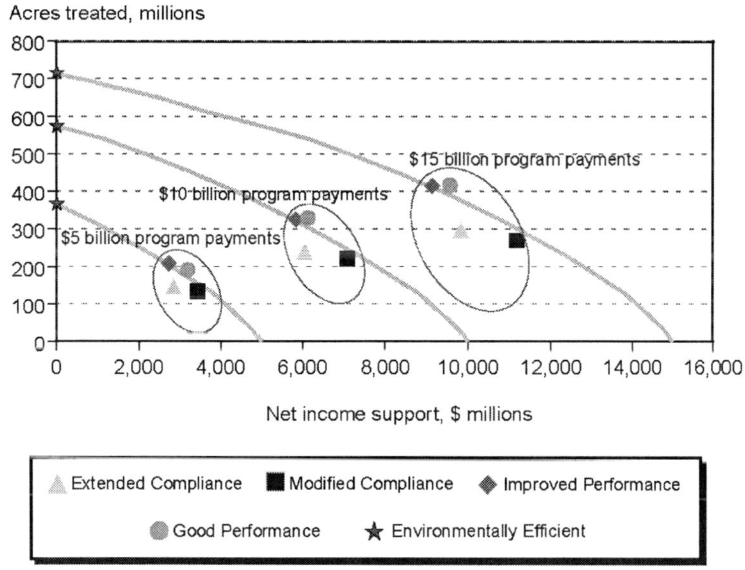

Source: USDA, Economic Research Service.

Figure 6. Efficiency matters: Trading treated acreage and income support in green payment program design.

Consider the *Improved Performance* scenario on the $10-billion frontier (the middle of the three curves). Of the $10 billion in producer payments, roughly $4 billion offsets conservation costs, yielding about 16 million environmental points and leaving $6 billion for income support. Because this scenario is already on the cost-effectiveness frontier, it would be impossible to increase environmental gain without reducing overall income support, and vice versa, while staying within the overall $10-billion budget. The same is true for the *Environmentally Efficient* scenario: if some of the $10 billion in payments were allocated to income support, overall environmental gain would decline.

Our other green payment scenarios are not on the cost-effectiveness frontier, indicating that more environmental gain, more income support, or both could be obtained without increasing overall program payments. The *Good Performance* scenario, in which "good actors" are eligible for payments even if they do not improve their environmental performance, is close to the cost-effectiveness frontier, delivering slightly less environmental gain and slightly more income support than the *Improved Performance* scenario. In this scenario, the decision to support producers who have achieved a relatively high level of environmental performance—even if they take no action to improve their performance—is achieved at the cost of a modest loss in environmental gain.

In contrast, the environmental compliance scenarios produce substantially less environmental gain than either environmental performance scenario. The *Extended Compliance* scenario, moreover, delivers very little additional overall income support. The *Modified Compliance* scenario does a bit better—it would yield the same or slightly more environmental gain than *Extended Compliance* while also producing a higher level of overall income support. In these scenarios, the decision to direct support toward the current recipients of farm commodity program payments comes at the cost of a substantial loss in environmental gain.

Differences in cost-effectiveness across our four scenarios are largely a function of three key determinants: 1) the broadness of eligibility requirements; 2) the effectiveness of payment incentives in encouraging the participation of producers who can deliver environmental benefits at low cost; and 3) the flexibility that producers have in responding to those incentives. In the environmental performance scenarios, broad eligibility ensures that producers who can deliver cost-effective environmental gains can participate. Payments reflect the potential for environmental gain, encouraging the participation of producers who can offer environmental gains that are large relative to the cost of obtaining these gains. Finally, because producers are free to decide which tracts of land they will offer for enrollment and which practices they will adopt on those tracts, they are free to offer only the land and practices for which the payment (which is proportional to environmental gain) exceeds economic costs.

By contrast, *Extended Compliance* offers only limited eligibility, payments that are unrelated to environmental gain, and no flexibility on environmental requirements. Producers must meet all environmental requirements or face loss of eligibility for payments. If producer economic cost varies widely across conservation treatments, individual producers may be able to make a cost-effective contribution toward some environmental objectives but not others. Some farms would elect to undertake these treatments because their overall payment would be larger than their overall economic cost of conservation improvements. Other farms would decide not to participate at all, given the level of payment they could receive and the cost they would incur. As a result, the overall economic cost of environmental gain is high.

Modified Compliance is more environmentally cost-effective than *Extended Compliance* because the opt-out provision offers producers both flexibility and the incentive to exercise that flexibility in a way that increases cost- effectiveness. Unlike *Extended Compliance, Modified Compliance* allows producers to opt out of some requirements, if they agree to a reduction in payment commensurate with the loss of environmental gain due to the opt- out. If the payment reduction is commensurate with the level of benefits forgone, producers will opt out only when the benefits of a given conservation treatment fail to outweigh the cost of the treatment. In other words, producers will opt out of treatments only when they are not cost-effective. Net income support to participating producers also increases (relative to the extended compliance scenario) because the reduction in payment is less than the reduction in economic cost. By allowing producers to focus on relatively cost-effective environmental gains, both environmental gain and income support can be increased in relation to *Extended Compliance.*

To what extent is higher cost effectiveness in the environmental performance scenarios driven by differences in eligibility versus differences in incentives? That is, are the environmental performance scenarios more cost effective simply because they can include a broader range of producers and land? To separate these effects, the environmental performance scenarios were re- estimated, restricting eligibility to recipients of existing income support and excluding payments for wildlife habitat-related treatments, because this is not required in the compliance scenarios. Figure 7 shows the cost-effectiveness curves for all four green payments scenarios but also shows curves for the *Improved Performance* and *Good Performance* scenarios under the assumption that eligibility is limited to the same producers and resource concerns that are eligible for payments in the environmental compliance scenarios. Although the change in assumptions does increase the average cost of environmental gain in the environmental performance scenarios, costs are still substantially lower than in either environmental compliance scenario. These results show that both broad eligibility and effective incentives are needed to obtain cost-effective environmental gain.

How does conservation cost-effectiveness affect the distribution of income support? Policymakers may face a difficult tradeoff between environmental gain and the *distribution* of income support. The same program design features that lead to cost-effective environmental gain also result in a distribution of net income support, across producers, that is quite different from that of existing commodity programs.

A key factor is broad eligibility. While cost-effectiveness is enhanced by ensuring that all producers who can make a cost-effective contribution are included, net income support is also spread much more broadly across the farm sector. Figure 8 shows the number of participating farms for all four scenarios against total program payments. For program payments of $3 billion or more (total over 5 years), the number of farms participating in either environmental performance scenario exceeds participation in either environmental compliance scenario. Because participation in environmental performance scenarios is relatively large, payments per farm are small when compared to per-farm payments received through the environmental compliance scenarios.

Environmental performance and environmental compliance scenarios also vary in distribution across the ERS combined farm typology, commodity specialization, and ERS farm resource regions. Figure 9 shows the distribution of net income support across the ERS combined farm typology for the green payment scenario. Commercial farms (with gross annual sales of more than $250,000) capture the largest share of net income support in every scenario, although their share is somewhat larger for the compliance scenarios, particularly *Modified Compliance.* Compared to payments that are distributed like existing direct payments, however, income support is lower because of conservation treatment costs.

While commercial farms receive the largest share of income support, the environmental performance scenarios tend to shift support toward intermediate and rural residence farms. Intermediate farms (gross sales of less than $250,000) capture the next largest share of income support. For the performance-based scenarios, intermediate farms receive almost as much overall income support as commercial farms, although they would receive less on a per-farm basis because there are more intermediate farms than commercial farms. Rural residence farms, which tend to be small and are typically operated by retirees or individuals with full-time off-farm jobs, receive the smallest share of total income support. Nonetheless, these producers receive as much net income support in the performance-based scenarios as they would from payments that mimic the existing direct payments program.

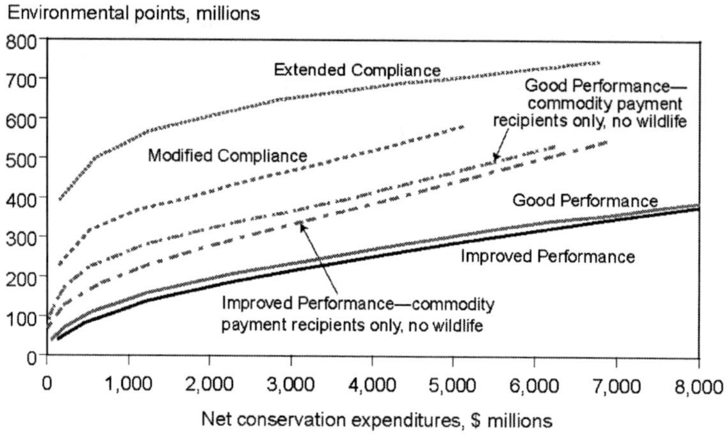

Note: Conservation expenditures don't reach $8 billion in some scenarios because overall payments are limited to $25 billion. The compliance-based scenarios yield larger farm income support than do performance-based scenarios, particularly when overall program payments are larger (see figure 4).
Source: USDA, Economic Research Service.

Figure 7. Net conservation expenditure per environmental point Environmental points, millions.

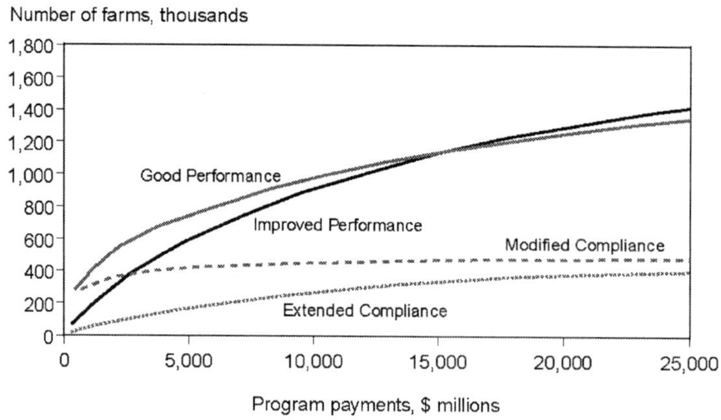

Source: USDA, Economic Research Service.

Figure 8. Number of participating farms by green payment scenario.

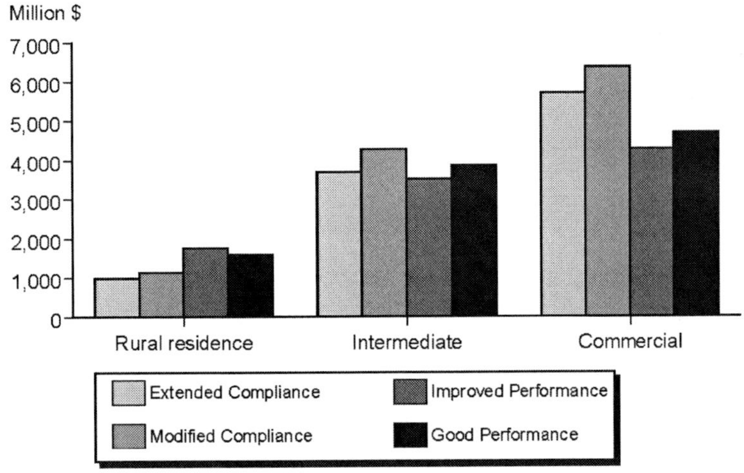

Source: USDA, Economic Research Service.

Figure 9. Net income support by scenario and combined typology.

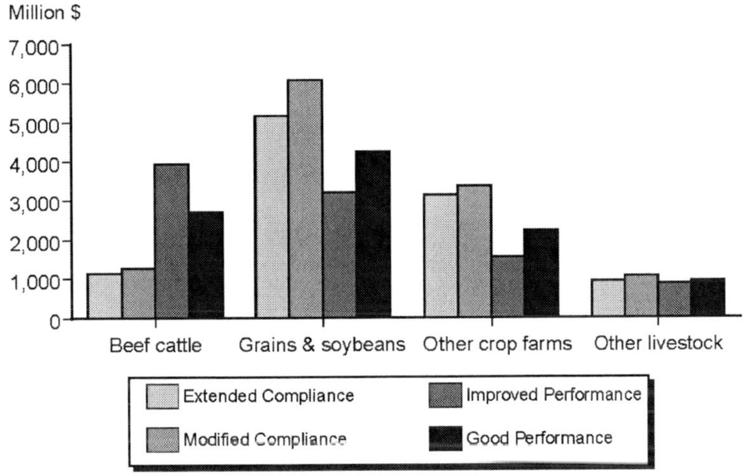

Source: USDA, Economic Research Service.

Figure 10. Net income support by scenario and primary commodity.

In terms of commodity specialization, beef producers do quite well in the environmental performance scenarios, despite the cost of conservation treatment (figure 10). Beef producers hold large acreages of grazing land, which is eligible for payments under *Improved Performance* and *Good Performance* but not for existing direct payments or under *Extended Compliance* and *Modified Compliance*. For crop producers, who tend to receive larger payments under existing income support programs, the situation is reversed: Net income support is larger for scenarios based on existing income support programs.

Regionally, income support tends to be large in the Heartland in all four scenarios, and particularly the two income support scenarios, because a large share of agricultural land, particularly cropland, is located there (figure 11). Payments in the *Improved Performance* and *Good Performance* scenarios tend to be more uniform across regions than for the compliance scenarios. Under *Improved Performance* and *Good Performance,* for example, regions like the Eastern Uplands and Basin and Range receive a substantial level of income support, even though they currently receive a relatively small share of direct payments. In the compliance scenarios, the Heartland and Prairie Gateway regions are favored because they receive a large share of existing direct payments.

Can the tradeoff between environmental cost-effectiveness and the distribution of income support be avoided? Our analysis indicates that only separate income support and conservation programs offer policy- makers full flexibility to tailor income support and conservation payments to maximize environmental gain and achieve a distribution of income support payments that matches that of the existing direct payment program (or any other distribution policymakers choose to implement). Because the environmentally efficient and pure income support scenarios are separate, the desired distribution of income support does constrain cost-effectiveness and vice versa. The mix of overall income support and environmental gain achieved with *Improved Performance* can also be achieved, at least in theory, with *Environmentally Efficient* payments and pure income support without any constraint on the distribution of income support. Even though this exact outcome would be difficult to achieve in reality, the additional constraint imposed by combining income and environmental payments would likely make it impossible.

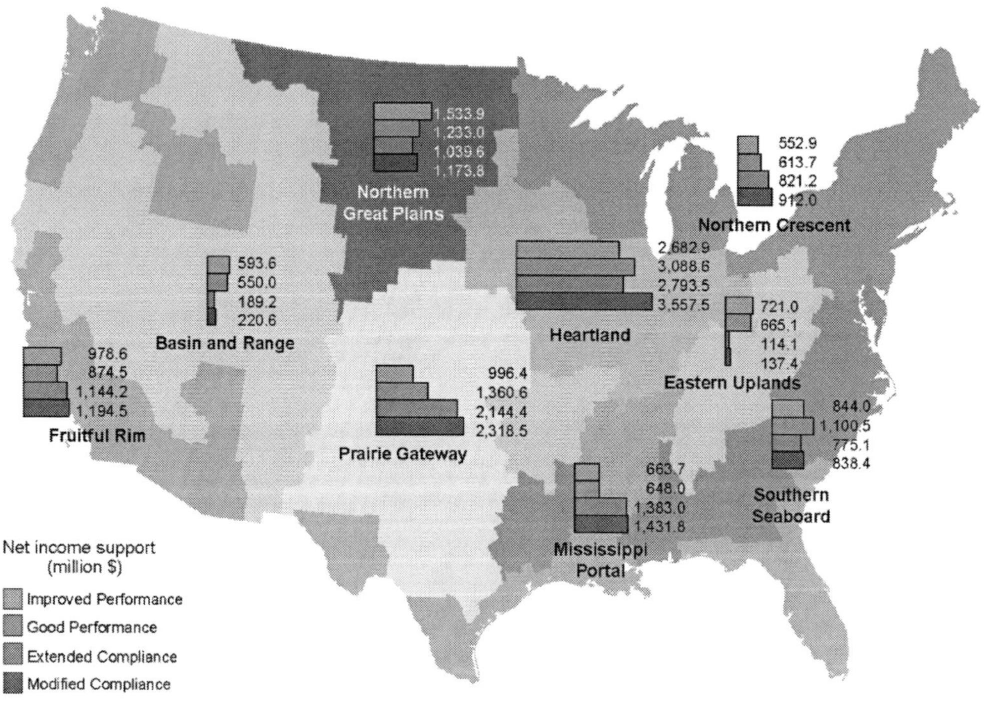

Source: USDA, Economic Research Service.

Figure 11. Net income support by scenario and ERS Resource Region, program payments $16 billion.

CONCLUSIONS

A green payment program that integrates commodity and conservation program objectives would provide both income support and environmental gain over a wide range of program designs. Our estimates indicate that 50 percent or more of payments could support farm income and that conservation treatment could be applied to several hundred million acres, given program budgets of $5 billion to $15 billion (total over 5 years). Nonetheless, the overall level of environmental gain and the distribution of income support can differ widely across green payment program designs.

If policymakers want the distribution of income support in a green payment program to approximate that of existing commodity program payments, they will need to accept less environmental gain than could be realized from a program designed from an environmental point of view. Opportunities for cost-effective conservation treatments do not always occur on farms that receive existing commodity program payments. A dollar spent on conservation in the environmental compliance scenarios would yield less environmental gain than it would if spent on conservation through one of the environmental performance scenarios. In general, green payments will be more environmentally cost effective when eligibility is broad, incentives encourage producers to take actions that yield high environmental benefit at low cost, and program participation requirements are flexible enough to allow producers to pursue these actions. Focusing on existing commodity program participants will exclude many producers who could deliver cost- effective environmental gain.

Of course, a program developed from a purely environmental point of view will also yield a dramatically different distribution of income support from that of existing direct payments. Although total income support is estimated to be similar in magnitude across all of the green payment scenarios we considered, those focused on environmental performance would spread income support more evenly across farms and regions. Crop farms would tend to realize less income support when compared to a program similar to the existing direct payment program while livestock farms (especially beef farms) would likely realize similar or larger support. Regions where small farms and livestock farms dominate (the Eastern Uplands and Basin and Range) would see a significant increase in support largely at the expense of regions that dominate existing farm commodity programs. Moreover, because participation is estimated to be greater in designs that focus on environmental performance, more producers benefit but income support per farm is much smaller than it would be for designs where eligibility for existing commodity programs is a starting point.

According to economic theory, accomplishing two goals in a cost-effective way will generally require two programs or policy instruments. That means that a single program of green payments will be less cost effective in achieving program goals than would separate programs for environmental gain and producer income support. As already shown, the combination of a program targeting environmental gain (similar to the *Environmentally Efficient* scenario) and another targeting income support (similar to the existing direct payments) can achieve a cost-effective outcome without restricting the distribution of income support. An environmental program similar, perhaps, to EQIP with competitive bidding, could be used to obtain environmental gain.

If policymakers decide to go forward without considering the existing distribution of commodity program payments, a green payments program could provide an alternative rationale for income support. Performance-based payments (analyzed in the *Improved Performance* and *Good Performance* scenarios) would allow producers to produce environmental "goods" for a price established by the Government. This approach would be somewhat similar to that of the Conservation Security Program (CSP), where some payments are currently based on reaching a threshold of environmental performance and other payments are based on environmental gain, measured by an environmental index. If policymakers view green payments as a longterm approach to income support and environmental gain, a mix of payments for new environmental gain and ongoing environmental stewardship may encourage producers to improve and, ultimately, maintain high environmental performance.

APPENDIX 1. ENVIRONMENTAL INDEX FOR GREEN PAYMENTS

Indexes are used widely in conservation programs to gauge the potential environmental gain from the application of proposed practices in a specific location. The environmental index described here is designed to capture the potential for environmental gain across a broad range of resource concerns. Resource concerns refers to resource attributes such as water quality, soil quality, and air quality, which are linked to agricultural production through physical effects such as soil erosion, nutrient runoff, and/or pesticide leaching.

Environmental indexes can be specified as indexes of potential environmental gain or environmental performance. For indexes that measure potential gain, a high value denotes higher potential for environmental damage (e.g., water quality damage from nutrient runoff) in the absence of conservation treatment (e.g., nutrient runoff control) or potential for lost opportunity to improve environmental performance in the absence of treatment (e.g., lost opportunity to enhance wildlife populations though habitat enhancement). Indexes of environmental performance are a mirror image: index values are high when there is little opportunity for environmental gain. When nutrient runoff has been controlled through nutrient management or other means, for example, the potential for further environmental gain is low, but environmental performance is high.

In our analysis, we use a performance-based index. Basic index derivations, however, are for a potential gain type index. The two types of indexes can be related as: $S_f = \max(I) \cdot I_f$. Where S_f is the performance-based index value for farm f, I_f is the potential environmental gain-based value for farm f, and $\max(I)$is the largest possible value of I.

For a given farm, the environmental index is an acre-weighted average of components that correspond to various treatments producers can apply to land in specific uses, such as water erosion control on land in crop production (the farm subscript is suppressed to avoid clutter):

$$I = \frac{\sum_k \sum_j A_{kj} I_{kj}}{\sum_k \sum_j A_{kj}} \tag{1}$$

Where

A_{kj} is the number of acres eligible for treatment k on land in use j;

I_{kj} is the index component representing the potential environmental damages or benefits that could be mitigated through application of treatment k on land in use j.

Each index component is the weighted sum of subcomponents representing the potential for damage from the physical effects (m) that could be addressed by applying treatment k on land in use j:

$$I_{kj} = \sum_{m \in k} w_{jm} N_{jm}$$

Where
m indexes physical effects;
w_{jm} is the index weight for land use j and physical effect m;
N_{jm} is the normalized index subcomponent for land use j and physical effect m;

The treatment index (k) is not used at the subcomponent level because each corresponds to a specific physical effect which, in turn, corresponds to only one treatment (although a treatment can correspond to more than one physical effect).

Finally, the basic building block of each index subcomponent (each N) is the relative damage (or benefit) estimate or RDE. Each RDE is the product of variables that describe (1) the intensity of the relevant physical effect (on the individual field) and (2) the potential for that physical effect to cause environmental damage:

$$RDE_{mj} = E_{mj} D_{mj}$$

Where
E_{mj} is the intensity of the physical effect m on land in use j;
D_{mj} is the potential damage associated with the physical effect m on land use j.
The index subcomponents are a normalization of the RDE, see page 33.

Intensity of Physical Effects (E)

The intensity variable generally measures the on-field risk for adverse physical effects such as soil erosion or nutrient runoff.

Wind Erosion: Intensity is measured as the average estimated excess wind erosion (wind erosion in excess of T) per acre for land that has excess erosion, by county. Estimates of wind erosion and the soil loss tolerance (T) are obtained from the National Resources Inventory (NRI). NRI is an area- based survey of land conducted by USDA's Natural Resources Conservation Service (NRCS) in cooperation with Iowa State University. NRI provides information on land use, land characteristics, and land condition (including estimated erosion rates), for about 800,000 points of non-Federal land across all U.S. counties, except those in Alaska.

Water Erosion: Intensity is measured as the average estimated excess water erosion (water erosion in excess of T) per acre for land that has excess erosion, by county. Estimates of water erosion and the soil loss tolerance (T) are obtained from the NRI.

Nutrient Management: Indexes that measure the risk of nitrogen and phosphorous runoff and nitrogen leaching to groundwater are developed using NRI and other data. See appendix 1 in Claassen et al. (2004) for details on the construction of the indexes.

Table A1.1. Intensity and damage variables, by index component and subcomponent.

Component (treatment (*k*) and land use (*j*))	Subcomponent (physical effect (*m*))	Intensity	Damage
Wind erosion (cropland)	Dust Soil productivity	Excess erosion, NRI[1] Excess erosion, NRI	$/ton of erosion, ERS $/ton of erosion, ERS
Water erosion (cropland)	Sediment Soil productivity	Excess erosion, NRI Excess erosion, NRI	$/ton of erosion, ERS $/ton of erosion, ERS NA[2]
Nutrient management (cropland)	Nitrogen leaching Nitrogen runoff Phosphorous runoff	NRI-based index NRI-based index NRI-based index	Transport to estuary, USGS SPARROW[3] NA
Pest management (cropland)	Pesticide leaching Pesticide runoff	Index of pesticide concentration in water leaching below root zone[4] Index of pesticide concentration in runoff[4]	NA NA
Grazing land health (grazing land)	Grazing land health	Index of potential productivity Improvement[5]	NA
Nutrient mgmt and riparian erosion (grazing land)	Nutrient runoff and sediment	Index of nonconfined animals/acre and stream density on grazing land	NA
Wildlife (cropland and grazing land)	Habitat restoration	Number of imperiled species in county[6]	NA

[1] National Resources Inventory (see text for description of NRI data).

[2] NA = data not available

[3] US Geological Survey, Spatially Referenced Regressions of Watershed Attributes. See http://water

[4] Goss et al.

[5] Atwood et al.

[6] Based on NatureServe natural heritage data, see www.NatureServe.org.

Source: USDA, Economic Research Service.

Pest Management: Data obtained from Goss et al., (1998) indicates the number of acres, by watershed (as defined by 8-digit hydrologic unit codes (HUC)), for which pesticide concentration exceeds human health standards in surface runoff and water leaching below the root zone. They also note the number of acres where runoff and leaching contain pesticides in concentrations that are 2 times, 3 times, 5 times, etc., up 25 times, the safe concentra-tion for human health. The intensity measure for pesticides is: $E = \sum_t tc_1$ where t is the number of times pesticide concentration in runoff or leaching exceeds the human health standard and c_t is the number of acres for a given t.

Grazing land health: The grazing measure is based on Atwood et al., (2005). Here pasture productivity is considered a proxy for good grass cover which controls erosion and active weed control which makes it more difficult for invasive species to take hold. The index is calculated as the difference between local (watershed) average forage yields and "high" forage yields (defined as the local mean yield plus two standard deviations of forage yield. In other words, the index measures difference between what is typically obtained versus what could be obtained though superior management.

Nutrient management and riparian erosion on grazing land: The index subcomponent value varies spatially based on the intensity of grazing (number of animal units per acre) and the density of streams relative to grazing land. By county, we estimate the number of grazing animal units and grazing land acreage from the Census of Agriculture. The U.S. Department of Commerce's U.S. Census Bureau conducted the Census of Agriculture for 156 years, 1840-1966. Starting with the 1997 Census of Agriculture, Congress moved that responsibility to USDA's National Agricultural Statistics Service, which collects data in a 5-year cycle for years ending in 2 and 7. Stream density is measured by the average distance of grazing land from perennial streams within counties using National Land Cover Data, National Hydrography Data, and the National Elevation Dataset. The raw index subcomponent is the ratio of stocking density to average distance of grazing land to streams. The higher the stocking density and the shorter the (average) distance to water, the higher the index value.

Wildlife: The wildlife component score is based largely on actions rather than location. Ten percent of index points are based on an intensity variable, defined as the number of potentially imperiled species in the county, as measured by NatureServe (see www.N-atureServe.org). The balance of wildlife points are awarded for taking action to improve habitat.

Potential Damages (D)

The damage variable is a measure of the potential for physical effects to cause damage to specific resource attributes such as water quality or soil productivity. These damages may be expressed in monetary terms, may be represented by proxies, or may be absent altogether.

Wind Erosion: Ribaudo et al., (1990) developed measures of the cost of reduced air quality due to particulate pollution caused by wind erosion. Wind-born dust costs include cleaning and maintenance of businesses and households, damage to non-farm machinery, and adverse effects on human health (Piper and Huzar, 1989). Cost per household is modeled as a function of the wind-erosion rate, income, and other household characteristics. The cost model is estimated using contingent valuation techniques and data from a survey of households in New Mexico. The cost model is applied to households west of the Mississippi River using Population Census data and wind erosion estimates. Results are aggregated across households within USDA Farm Production Regions. Damage (benefit) estimates are provided per ton of soil eroded (conserved).

Reductions in soil erosion (for wind or water erosion) will increase the future productivity of farmland. Ribaudo et al., (1990) used yield losses and production-cost increases due to erosion estimated using the Erosion Productivity Impact Model (Williams et al., 1985). The economic value of the gain in productivity is the net current value of the increase in productivity resulting from a marginal reduction in soil erosion.

Water Erosion: The change in consumer surplus associated with water-based recreation due to a change in soil erosion within a watershed is based on Feather and Hellerstein (1997) and Feather et al., 1999). Demand for water- based recreation is estimated using behavioral data from the 1992 National Survey of Recreation and the Environment (NSRE) and soil erosion estimates from the NRI. Demand is modeled as a function of the individual's characteristics, travel costs, erosion levels and other environmental factors. Across the 2,111 HUCs, a 1-ton erosion reduction can increase societal benefits of water-based recreation from 0 to $8.81.

Hansen et al. (2002) estimate the water quality damages of soil erosion within a HUC based on the cost of sediment to downstream navigation. They develop a hydrologic model that accounts for the hydrology and the subsequent flow of sediment within and across watersheds. Their hydrologic model links erosion within a watershed to the downstream cost of dredging harbors and shipping channels. The hydrologic data are from the Environmental Protection Agency's River Reach File, which shows interconnections among 3.2 million miles of streams. Estimates of agricultural erosion by HUC are based on data from the NRI. Dredging-cost data are from the U.S. Army Corps of Engineers. Results show that, across HUCs, a 1-ton reduction in soil erosion can reduce dredging costs from zero to $5.00.

A range of other water quality benefits are obtained from Ribaudo (1990) including:

- *Commercial fishing* benefits, which result from reduced sediment loads in coastal estuaries that serve as breeding grounds for many species;
- *Flooding*-related benefits derive from reduced cost of flood clean up due to reduced sediment concentrations in flood waters;
- *Water conveyance* benefits result from reduced cost of removing sediment from water conveyance facilities, primarily drainage ditches and irrigation canals;
- *Water treatment* benefits are the result of lower water treatment costs due to reduced sediment loads;
- *Municipal and industrial* benefits are due to reduced damage to water-use equipment from minerals, salts, and other materials associated with soil erosion;
- *Steam-electric power plants* that rely on water-cooling benefit from reduced sediment through reduced wear on facilities.

Nutrient Management (nitrogen runoff): Water quality damage due to nitrogen can occur anywhere, but is more common in coastal estuaries where nitrogen, rather than phosphorous, is most often the limiting nutrient in excess algae growth and eutrophication. The likelihood of nitrogen transport to coastal areas is used as a proxy for potential water quality damage. Transport coefficients, which represent an estimate of the proportion of nitrogen runoff that is transported to the coast, were drawn from the SPARROW model developed by U.S. Geological Survey researchers (Smith, Schwartz, and Alexander, 1997).

Normalizing Index Subcomponents

Relative Damage Estimates (RDEs) are normalized to the unit interval:

$$N_{jm} = \frac{\max(RDE_{jm}, P95(RDE_{jm})) - \min(RDE_{jm})}{P95(RDE_{jm}) - \min(RDE_{jm})}.$$

where RDE_{jm} indicates a farm-specific value and $\min(RDE_{jm})$ and $P95(RDE_{jm})$ indicate the minimum and 95th percentile values across all farms, respectively. The 95th percentile is used, rather than the maximum, to prevent outliers in the data from depressing index scores for other farms.

A special consideration applies to the soil productivity component. Because loss of soil depth to wind or water erosion can cause productivity damage, the soil productivity weight is the same for wind and water erosion and is designed to capture the full value of soil loss from both sources. Including independently normalized intensity/damage terms for each would result in double-counting for soil productivity in relation to the normalized intensity/ damage terms for sediment runoff or windblown dust. To correct for this possibility, the soil productivity terms are normalized as:

$$0 \leq \varphi N_{crop,water} + (1-\varphi)N_{crop,wind} \leq 1,$$

$$\text{where } \varphi = \frac{RDE_{crop,water}}{RED_{crop,water} + RDE_{crop,wind}}.$$

These additional weighting factors are included with the normalized subcomponents for water and wind erosion.

Table A1.2. Initial index weights.

Component (treatment (k))	Subcomponent (physical effect (m))	Weights		
		Nonirrigated cropland	Irrigated cropland	Grazing land
Wind erosion	Dust Soil productivity	.03 .10	.03 .10	-- --
Water erosion	Sediment Soil productivity	.07 .10	.07 .10	-- --
Nutrient management	Nitrogen leaching Nitrogen runoff Phosphorous runoff	.04 .04 .04	.04 .04 .04	-- -- --
Pest management	Pesticide leaching Pesticide runoff	.04 .04	.04 .04	-- --
Grazing land health	Grazing land health	--	--	.06
Nutrient management and riparian erosion	Nutrient and sediment grazing land	--	--	.14
Wildlife	Habitat restoration	.20	.20	.20
Totals		.70	.70	.40

Source: USDA, Economic Research Service.

Index Weights

Initial weights are based roughly on the EBI, with points added for the inclusion of grazing land in the model:

- Roughly one-third of points are for soil erosion on cropland. Points are given for potential of erosion control to reduce dust (improve air quality), preserve soil productivity and reduce sediment loads to water.
- Another third are for other water quality-related treatments, including nutrient management and pest management on cropland, nutrient management and riparian erosion on grazing land, and grazing land health.
- Remaining points are for wildlife habitat enhancement (split evenly among all 3 land types).

The weights that are actually used in the model are given in table A1 .2.

APPENDIX 2. ACRES ELIGIBLE FOR TREATMENT

For most physical effects modeled, land is eligible for treatment if the severity of one or more physical effects exceeds a predetermined threshold level. For example, land is eligible for soil conservation treatment where soil erosion exceeds the soil loss tolerance or "T" level.[5] Other physical effects are listed in table 1 in the main text. This appendix describes the data sources and our approach to linking data on conservation treatment needs to ARMS farms.

Eligibility doesn't mean that the producer will undertake any specific treatment or that he/she will receive a payment for it, only that the treatment is appropriate for the land in question. Decisions about whether land will be treated depend on whether the producer is willing to accept the level of payment offered in exchange for treating the land.

Total acreage eligible for treatment is estimated to be 829 million acres (table A2.1). Because one actual acre can be eligible for more than one treatment, some acres are counted more than once. For example, if a single acre of nonirrigated cropland is eligible for nutrient management and soil erosion control, it would be counted twice in table A2.1. Thus, total actual acreage that could receive one or more treatments is less than 829 million acres.

Data and Estimation

Basic land use data, by farm, is from ARMS, phase 3, for 2002. The 2002 survey provides a full accounting of all land uses (e.g., crop production, grazing) on each farm in the survey. Data on soil erosion, potential pesticide loss, and other physical effects, however, is not collected with the ARMS data. Other data are used to estimate the proportion of acres needing treatment for a specific physical effect in the county or watershed that includes the farm (table A2.2). That information is used together with farm-specific acreages to estimate acres eligible for treatment:

$$A_{kj} = A_j P_{kj}$$

where:

A_{kj} is farm-specific acreage estimated to need treatment k on land in use j;

A_j is farm-specific acreage in land use j; and,

P_{kj} is the proportion of acres in need of treatment k on land in use j in the surrounding area.

Soil erosion (wind and water): Cropland acres are eligible for soil conservation treatment if estimated average annual erosion (wind or water) exceeds the soil loss tolerance (T) value for the land. The proportion of acres eligible for treatment, by county, is estimated from the NRI, which provides estimates of soil erosion rates, T-values, and acreage for all cropland sample points.

Nutrient runoff and leaching: All cropland acres are eligible for treatment for nutrient runoff and leaching. This assumption reflects the fact that excess nutrient balances and nutrient loss to water are widespread problems and thresholds for nutrient treatment are not clearly defined.

Table A2.1. Treatable acres by land use and treatment.

Land use	Treatment	Acres (millions)[1]
Nonirrigated cropland	Water erosion control	52.00
Nonirrigated cropland	Wind erosion control	26.77
Nonirrigated cropland	Wind and water erosion control	2.51
Nonirrigated cropland	Nutrient management	280.48
Nonirrigated cropland	Pest management	109.39
Nonirrigated cropland	Wildlife habitat enhancement	4.59
Irrigated cropland	Water erosion control	2.03
Irrigated cropland	Wind erosion control	4.31
Irrigated cropland	Wind and water erosion control	0.28
Irrigated cropland	Nutrient management	49.86
Irrigated cropland	Pest management	17.27
Irrigated cropland	Wildlife habitat enhancement	0.47
Grazing land	Grazing land health enhancement	218.91
Grazing land	Nutrient management and riparian erosion control	44.62
Grazing land	Wildlife habitat enhancement	15.78
All grazing land	All	279.32
All irrigated cropland	All	74.22
All nonirrigated cropland	All	475.74
All treatable acres	All	829.27

[1] Total acres treated will be less than 829 million because some actual acres are eligible for more than one treatment.

Source: USDA, Economic Research Service.

Table A2.2. Definition of acres needing treatment, by physical effect.

Treatment	Relevant ARMS acreage	Proportion of acres (county or HUC) needing treatment	Data Source
Wind erosion	Cropland[1]	Proportion of acres with wind erosion > T^2	NRI
Water erosion	Cropland	Proportion of acres with water erosion > T	NRI
Nutrient management	Cropland	All acres assumed to need treatment	--
Pesticide management	Cropland	Proportion of acres for which pesticide concentration in water leaving field is estimated to exceed human health threshold	Goss et al.
Grazing land health	Grazing land	Grazing acres cited as needing treatment for grazing land health	WLA
Nutrient mgmt and riparian erosion	Grazing land	Grazing acres cited as needing treatment for nutrient-related concerns or soil erosion	WLA
Habitat restoration/ wildlife management	Cropland	Cropland acres cited as needing treatment for wildlife concerns	WLA
	Grazing land	Grazing acres cited as needing treatment for wildlife concerns	WLA

[1] Nonirrigated and irrigated cropland are treated separately in the analysis.

[2] While the T factor is based on the potential for productivity loss and not water quality, per se, it is reasonable to assume that full treatment has been achieved when soil erosion is reduced to T.

Source: USDA, Economic Research Service.

Pesticide runoff and leaching: Cropland acres are eligible for pesticide management if water leaving the field, either through surface or subsurface flows, is estimated to contain pesticides in concentrations that exceed standards for the protection of human health. Estimates of the number of acres by watershed (8-digit HUC) are obtained from Goss, et al.

Cropland habitat acres: Acreage eligible for habitat enhancement, by county, is from the NRCS Workload Assessment (WLA). WLA provides the acreage of various land types (e.g., cropland, grazing land) where conservation treatment for various physical effects would be appropriate. WLA estimates are based on the expert judgment of the local NRCS staff that assists farmers in determining conservation needs, devising conservation plans, and facilitating conservation practice implementation. To obtain the overall proportion of cropland acres eligible for habitat enhancement, WLA acreages are divided by the appropriate acreage (nonirrigated or irrigated cropland) obtained from the 2002 Census of Agriculture.

Grazing land grass cover, nutrient management, and habitat acres. On grazing land, county-wide estimates of acreage eligible for treatment for poor grass cover, nutrient-related issues, or habitat enhancement are from the WLA. To obtain proportions, WLA acreages are divided by grazing land acreage obtained from the 2002 Census of Agriculture.

On land that could be treated for both water and wind erosion, the cost of treating both is often less the sum of the separate treatments. That's because some practices can used for control of both water and wind erosion. A single practice can do both because wind and water erosion occur under different conditions and often at different times of the year. To avoid double counting the cost of these practices, 3 acreage variables are needed: total acres eligible for water erosion treatment, total acres eligible for wind erosion treatment, and acres eligible for both. Fortunately, the NRI data can provide these estimated. See appendix 3 for more detail on how these acreages are used in estimating treatment cost.

There is also overlap between treatment for N runoff, P runoff, and N leaching. All are addressed by nutrient management (NRC S practice code 590) but could be made up of differing components of the nutrient management practices. Although the plan is typically based on a single nutrient, all nutrients must be used at appropriate rates. So, when a producer undertakes nutrient management, we assume that all nutrients are applied at appropriate rates and otherwise managed in a way consistent with the spirit of the nutrient management plan.

APPENDIX 3. PRODUCER ECONOMIC COST/WILLINGNESS TO ACCEPT (WTA)

A producers' willingness to accept (WTA) is the minimum payment he or she would accept in exchange for taking a specific action, such as undertaking a given conservation treatment. WTA depends on a range of factors including (but not necessarily limited to) out-of-pocket costs (or savings), changes in production (positive or negative), changes in production risk, and the level of management skill required for successful implementation. For a given offer of payment in exchange for taking a specified action, a producer will accept the payment (apply for the programs in question) if the payment is equal to or larger than the producer's WTA.

In the text, we often refer to WTA as "economic cost" because WTA represents a producer's cost of applying conservation practices is a very broad sense of the term cost. Economic cost includes everything farmers must pay for (out-of-pocket expenses), but also includes things that are real but more difficult to estimate with precision. For example, producers may need to manage fertilizer applications more carefully to comply with terms of a nutrient management plan. This "hassle factor" is a real cost to the producer, but is hard to value in monetary terms.

In general, WTA is private information. Producers know what payment they would be willing to accept while policymakers and program managers do not. For model development, EQIP payments are used as a proxy for WTA. EQIP participants have shown that they are willing to accept the contract payment in exchange for adopting or installing practices specified in the contract. The EQIP payment is only a proxy because participants may also have been willing to accept lower payments. Moreover, producers who are not EQIP participants may or may not be willing to accept this same level of payment in exchange for taking similar actions. In the absence of other data, however, we assume that EQIP payments can serve as a reasonable proxy for WTA.

Information on the distribution of WTA across farms is given in table A3.1. For each land use and treatment combination, we provide a national average per acre WTA as well as the 5th, 25th, 50th, 75th, and 95th percentiles. For example, WTA for water erosion control on nonirrigated cropland averaged \$58 per acre, but 5 percent of acres that are eligible for this treatment can be treated for \$16.28 or less, 25 percent of acres can be treated for \$29.79 or less, and so on. Note that even though average per-acre WTA is higher for grazing land health (\$81.56) than for water erosion on non-irrigated cropland (\$58.04), a substantial share of grazing land could be treated for grazing land health at a relatively low cost: 25 percent of eligible acres have WTA estimates of \$17.12 per acre or less while the same percentage of land eligible for water erosion control has WTA of \$29.79 or less.

Data and Estimation

We estimate WTA as the average payment per treated acre for a suite of practices typically employed in addressing a specific physical effect in a given watershed (8-digit HUC). For example, conservation tillage, terraces, and contour farming, among other practices, are widely used to control the physical effect sheet and rill erosion (details below). By focusing on specific areas, our WTA estimates reflect both the frequency with which specific practices are used and the out-of-pocket cost, etc., of applying those practices in that area. The 8-digit HUC was selected as the unit area because they are large enough to contain a substantial number of observations but small enough to capture both spatial diversity in the exact mix of practices used to treat a given physical effect and variation in the payment needed to leverage the application of these practices. In areas where data are sparse, some WTA estimates are calculated for larger watersheds.

Table A3.1. Distribution of willingness to accept payment (WTA) across modeled farms.

Land use	Treatment	Percentiles					Average
		5	25	50	75	95	
				Dollars per acre			
Nonirrigated crop	Water erosion control	16.28	29.79	46.76	75.02	137.20	58.04
Nonirrigated crop	Wind erosion control	7.39	19.73	26.82	38.77	62.27	30.40
Nonirrigated crop	Nutrient management	3.28	9.11	12.90	16.18	26.98	13.62
Nonirrigated crop	Pest management	4.75	10.37	12.52	17.02	58.49	20.02
Nonirrigated crop	Habitat restore/enhance	4.20	12.73	16.22	27.47	55.73	22.93
Irrigated cropland	Water erosion control	7.48	23.29	35.81	55.32	107.63	44.17
Irrigated cropland	Wind erosion control	3.61	16.17	27.27	40.36	61.70	28.33
Irrigated cropland	Nutrient management	2.66	7.15	13.13	16.37	41.55	14.91
Irrigated cropland	Pesticide management	3.37	9.24	13.33	31.99	75.64	26.91
Irrigated cropland	Habitat restore/enhance	2.94	8.74	12.42	12.73	46.83	13.91
Grazing land	Grazing land health	5.82	17.12	49.12	105.45	281.75	81.56
Grazing land	Nutrient management and riparian erosion control	7.21	20.03	36.77	93.33	279.79	79.27
Grazing land	Habitat restore/enhance	2.81	7.11	19.59	40.56	80.65	29.08

Source: USDA, Economic Research Service.

A first step in calculating estimates of WTA is to identify practices that could be used to treat specific physical effects. Table 1 in the main text shows how treatments are linked to land use, physical effects, and resource concerns.

Cropland: Table A3.2 shows which practices are assigned to each treatments to estimate WTA. These groupings are drawn largely from Atwood, et al., 2003 and Atwood, et al., 2005. Grouping practices this way is not meant to infer that every farm would use every practice on every acre when addressing a specific physical effect. Rather, the WTA estimate for a given practice group in a given HUC reflects the frequency with which each practice within the group is used and local cost of applying these practices. For example, if conservation tillage is used frequently for soil erosion control in a given HUC while seasonal residue management is not used, the cost per acre will reflect the fact that conservation tillage is locally adapted while seasonal residue management is not.

The calculation of WTA must also consider whether individual practices can simultaneously treat more than one physical effect. Many soil erosion control practices can reduce both wind and water erosion. Because wind and water erosion occur under different conditions, the same practice can be effective for both. Water erosion occurs during storm events when wind erosion is not a hazard, while wind erosion occurs when the soil surface is dry when water erosion is not a hazard. If both types of erosion control are needed, simply adding the estimated WTA for water and wind erosion control could lead to double-counting because some practices, included in estimates for both wind and water erosion control, would be applied only once. To account for this possibility, we define practices that are commonly used to address both wind and water erosion and the cost-savings if both are addressed on the same land.

Grazing land: A two-level screen is used to identify practices for grazing land treatments. First, we establish a set of practices that are commonly used on grazing land (table A3.3). Our list is drawn largely from Atwood et al., 2003 and Atwood et al., 2005 and augmented by information drawn from EQIP contract data (table A3.3). We assume that grazing land health is addressed by one of the practices on our list whenever it is associated with one of these NRCS classification codes[6]:

- Excessive erosion (PG1);
- Invasion of noxious weeds (PG2);
- Invasion of woody species (PG3);
- Other grazing land health issues (PG4);
- Loss of plant diversity – declining species (PP1);
- Plants not adapted to site (PP3);
- Insufficient water supply for livestock (WQ6).

For nutrient management and riparian erosion control, a similar set of practices is used (there is significant overlap), but practices are counted only when they appear with these NRCS classification codes:

- Animal waste, organics, and pathogens (WS2);
- Loss of riparian vegetation (WS6);
- Stream bank and shoreline erosion (WS8);
- Loss or degradation of riparian vegetation (PP2);
- Stream bank and shoreline erosion (PP4).

Because we cannot determine whether grazing land health concerns occur on the same acres as water quality concerns, we do not estimate payments to practices that address both. We do, however, exclude nutrient management practices where they are paired with livestock practices that involve waste handling structures and equipment typically found on large farms with confined animals. We exclude these contracts because they are more likely to reflect nutrient management costs on confined animal feeding operations (which is not our focus) than on grazing land (which is our focus).

Wildlife: We assume that wildlife habitat on working agricultural lands will be addressed primarily through use of Upland Wildlife Habitat Management (NRCS practice code 645) and associated practices (listed in the national practice standard for Upland Wildlife Habitat Management, see table A3.4). When these associated practices occur in conjunction with any habitat- related NRCS classification code, they are used in the estimation of WTA for wildlife-related treatment. Some practices are assumed to be used only on grazing land acres (e.g., range planting), while we assume that others could be used on either cropland or grazing land (see table A3.4). Again, because we cannot determine whether wildlife concerns occur on the same acres as other physical effects, we do not estimate payments to practices that address wildlife in conjunction with other physical effects or resource concerns.

EQIP contract data for 2003-05 are used. All dollar amounts are adjusted to the beginning of fiscal year 2007 (October 1, 2006 through September 30, 2007) using the Gross Domestic Product (GDP) implicit price deflator. We also treat structural or vegetative practices differently from management practices due to differences in implementation. For structural and vegetative practices, EQIP payments are in the form of cost-sharing on actual (onetime) installation costs. For management practices, which must be re-applied each year, EQIP participants receive annual incentive payments for a 3-year period. These payments are designed to smooth the transition to the use of new management practices. Our estimate of WTA for specific practices is based on the cost-share payment to structural or vegetative practices, or the net present value of 3 years' worth of incentive payments on management practices. Discounting future costs at a rate of 7 percent, the net present value of the transition payment is equal to the annual payment times 2.62.

EQIP contract data indicates which practices have been used on a specific tract of land, how those practices were classified by NRCS, and the total number of acres treated (in all tracts associated with a given contract). Tract size is estimated as the total treated acres for the entire contract divided by the total number of tracts listed in the contract. Using this data, we calculate the average payment per acre for treating a specific physical effect as:

$$c_{kj}^{s} = \frac{\sum_{h\in k}\sum_{g}\sum_{t} x_{hgjt}^{s}}{\sum_{g} z_{g}} \text{ for structural/vegetative practices and } c_{kj}^{m} = \frac{\sum_{h\in k}\sum_{g}\sum_{t} d x_{hgjt}^{m}}{\sum_{g} z_{g}}$$

for management practices where

x_{hgjt}^{s} is expenditure for structural or vegetative practice h on tract g in use j at time t (already adjusted to 2007 dollars);

$x^{m}{}_{hgjt}$ is expenditure for management practice h on tract g in use j at time t (already adjusted to 2007 dollars);

z_g is acreage in tract g;

d is a discount factor $d = r^{-1}(1-(1+r)^{-T}$ where r is the interest rate and T is the number of years the payment is made (d=2.62 for 7 percent discount rate over 3 years).

Note that location subscripts are suppressed to avoid clutter.

Formally, then, our estimate of WTA for treatment k on land in use j can be written as:

$$a_{kj}c_{kj} = a_{kj}(c_{kj}^{s} + c_{kj}^{m})$$

Table A3.2. Practices for cropland.

Code	Management	Practice name
Sheet and rill erosion only		
311	no	Alley cropping
330	yes	Contour farming
331	yes	Contour Orchard and other fruit area
386	no	Field border
410	no	Grade stabilization structure
585	yes	Contour strip-cropping
600	no	Terrace
638	no	Water and sediment control
716	yes	Anion polyacrylamide (PAM) erosion control.
741	no	Vegetative buffer strips
Wind erosion only		
380	no	Windbreak/shelterbelt establishment
392	no	Field windbreak
422A	no	Herbaceous wind barriers
422	no	Hedgerow planting
589A		yes Cross-wind ridges
589B	yes	Cross wind strip-cropping
589C	yes	Cross wind trap strips
609	yes	Surface roughening
612	no	Tree/shrub establishment
650	no	Windbreak/shelterbelt renovation
704	no	Agroforestry planting
Both sheet & rill and wind erosion		
328	yes	Conservation crop rotation
329A	yes	Residue management, no-till and strip till
329B	yes	Residue management, mulch till
329C	yes	Residue management, ridge till
340	yes	Cover crop
342	no	Critical area planting
344	yes	Residue management, seasonal
586	yes	Strip-cropping
758	yes	Strip–intercropping
Nutrient runoff/ leaching only		
590	yes	Nutrient management
Pesticide runoff/ leaching only		
595	yes	Pest management

Source: USDA Economic Research Service.

Table A3.3. Practices for grazing land.

Code	Management	Practice name
Grazing land health only		
314	no	Brush management
338	yes	Prescribed burning
380	no	Windbreak/shelterbelt establishment
460	no	Land clearing
510	yes	Pasture and hayland management
512	no	Pasture and hay planting
528A	yes	Prescribed grazing
548	yes	Grazing land mechanical treatment
550	no	Range planting
612	no	Tree/shrub establishment
650	no	Windbreak/shelterbelt renovation
Water quality only		
590	yes	Nutrient management
Grazing land health and water quality		
382	no	Fence
472	no	Use exclusion
561	no	Heavy use area protection
574	no	Spring development
575	no	Animal trails and walkways
578	no	Stream crossing
614	no	Trough or tank
762	yes	Planned grazing system

Source: USDA Economic Research Service.
Economic Research Service/USDA

Table A3.4. Practices for wildlife management.

Code	Management	Practice name
Practices for use on grazing land		
338	yes	Prescribed burning
472	no	Use exclusion
512	no	Pasture and hay planting
528A	yes	Prescribed grazing
550	no	Range planting
612	no	Tree/shrub establishment
Practices for use on any land		
390	no	Riparian herbaceous cover
511	yes	Forage harvest management
643	no	Restoration of rare/declining habitat
645	yes	Upland wildlife management
647	no	Early successional habitat development
648	no	Wildlife watering facility

Source: USDA Economic Research Service.

where a_{kj} is the number of treated acres. When practices overlap treatments (only between wind and water erosion) we define WTA as:

$$a_{kj}c_{kj} + a_{k'j}c_{k'j} - a_{kk'j}c_{kk'j}$$

$c_{kk'j}$ is the average EQIP expenditure, per treated acre, for practices in both group k and group k' on land use j;

$a_{kk'j}$ is the acreage that needs treatment practices in groups k and k' on land use j;

Again, note that estimates vary by HUC, but location subscripts are suppressed to avoid clutter.

APPENDIX 4. PRODUCER DECISION RULES

Because participation is voluntary, producers respond to the offer of green payments by deciding whether and how to participate. We assume that producers participate only when the overall payment exceeds WTA[7] plus a minimum return that accounts for transaction costs. We assume transaction costs of $200 per farm in our analysis and test the sensitivity of results to changes in this threshold. When more than one participation option meets participation decision criteria, farms are assumed to select the one that yields the largest net return (payment less WTA).

For scenarios that use the environmental index including *Improved Performance, Good Performance,* and *Modified Compliance*, program offers and decision rules are illustrated graphically. In response to any performance-based incentive, producers will first choose the treatment that yields the largest number of environmental points (most performance) per dollar of WTA. In figure A4.1, that treatment is illustrated by point A. Next, the producer would consider treatment for the physical effect with the next- best combination of points and WTA (point B in figure A4.1), and so on. The points form a curve that can be thought of as the producers "supply" of conservation treatments.

This supply curve, combined with information on payments, determines which conservation treatments a producer will undertake. In the *Improved Performance* scenario, producers are paid on a per-point basis, starting at the producer's own baseline (pre-program environmental performance). In figure A4.2, the slope of the payment line represents the payment rate (dollars per environmental point). The distance between the payment line and the horizontal axis represents the payment amount and the distance between the points on the supply curve and the payment line represents the producer's net gain or income support, that is, the amount paid over and above the WTA for adopting environmentally sound practices. Given our assumption about producer behavior, they will choose the option with the largest net return.

In the *Good Performance* scenario, producers receive payments based on their performance relative an environmental hurdle rate rather than a farm- specific baseline (figure A4.3). If a given farm's baseline is above the hurdle (that is, the baseline is to the right of the hurdle as measured on the horizontal axis in figure A4.3), the producer receives a payment larger than could be earned under the *Improved Performance* scenario. Note that other producers, who have baselines below the hurdle rate (the baseline is to the left of the hurdle rate on the horizontal axis in figure A4.3), would be eligible for payments that are less than they would receive from the *Improved Performance* scenario. These producers may still participate, although their net return (income support) will be less than it would have been under *Improved Performance*. Thus, income support in *Good Performance* is focused on those producers who have already achieved a high level of environmental performance.

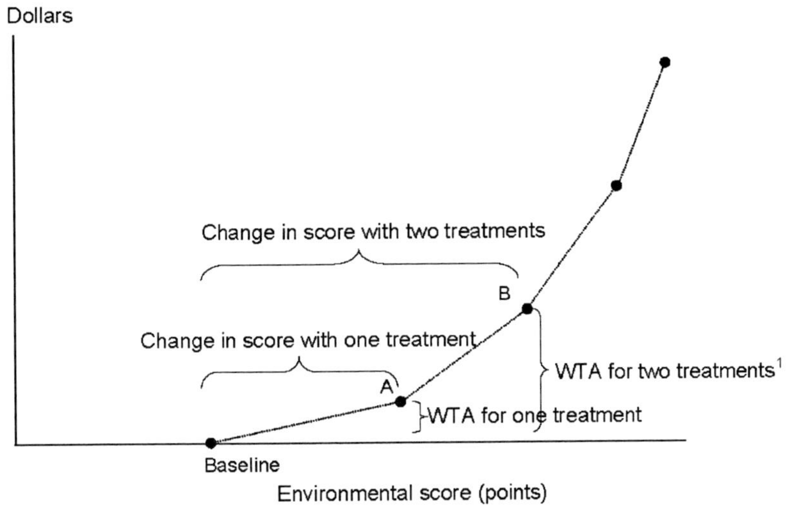

[1] WTA is defined as a producer's willingness to accept—the minimum payment he or she would accept in exchange for taking a specific action (e.g., undertaking a given conservation treatment).

Source: USDA, Economic Research Service.

Figure A4.1. Farm-level "supply" of conservation treatments.

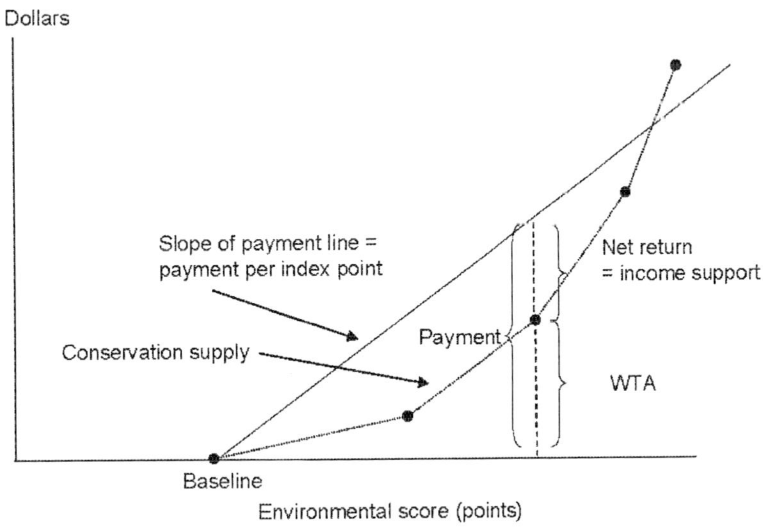

Source: USDA, Economic Research Service.

Figure A4.2. Decision rule for the *Improved Performance* scenario.

Finally, the *Modified Compliance* decision rule can also be depicted graphically as well (figure A4.4). The maximum payment is similar to the existing direct payments. Producers can receive 20 percent of their maximum payment even if they take no action to address additional environmental concerns. Producers receive 100 percent of their maximum payment if they address all remaining concerns. If the producer addresses only a portion of remaining concerns, the payment is pro-rated according to the proportion of remaining concerns that are addressed (as measured by the environmental index).

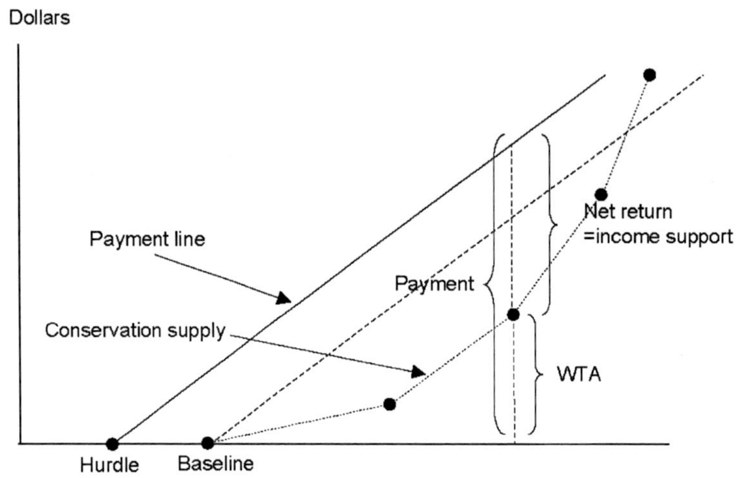

Source: USDA, Economic Research Service.

Figure A4.3. Decision rule for *Good Performance* scenario when producer baseline exceeds hurdle rate.

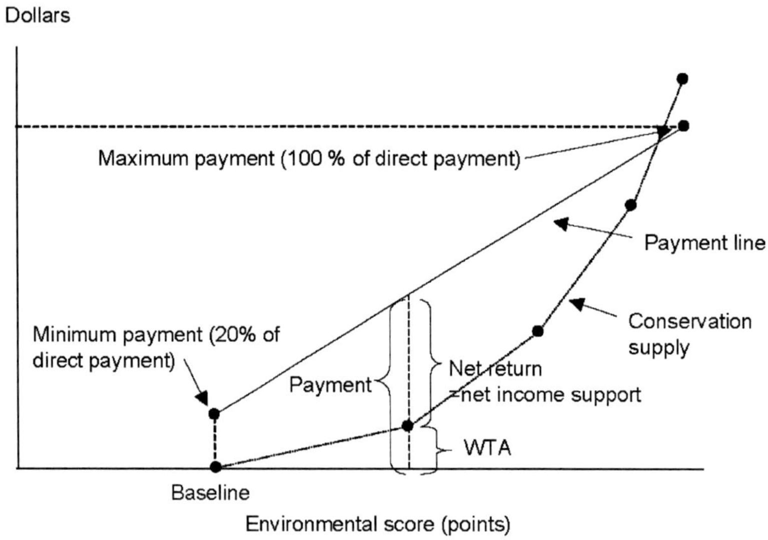

Source: USDA, Economic Research Service.

Figure A4.4. Decision rule for *Modified Compliance* scenario.

APPENDIX 5. SENSITIVITY ANALYSIS

We test model sensitivity to a number of economic and policy parameters selected for the main analysis:

- The minimum net return (which stands in for transactions cost);
- Environmental index weights; and
- Hurdle rates selected for the *Good Performance* scenario.

For the minimum net return and environmental index weights analyses, we use the *Improved Performance* scenario. In each of these sensitivity analyses, we attempt to hold overall program payments constant at about $15 billion over 5 years ($3 billion per year).

Sensitivity of Minimum Net Return

Producers do not have an economic incentive to participate in a green payments program unless participation results in a minimum return that at least covers transaction costs. These costs could include time and travel involved in filling out applications, verifying existing environmental performance, and working with conservation technicians to develop plans for additional conservation. In our initial analysis, a minimum return of $200 per farm was required to trigger program participation. We tested the sensitivity of model results to higher minimums over an extremely wide range, up to nearly $8,000 per farm.

Figure A5.1 shows that higher minimum returns could substantially reduce the number of farms that are estimated to participate, but would make much less difference in terms of treated acreage, environmental points, and net income support. When the minimum net return is $1,000, estimated participation (farms) drops by 40 percent, while change in other measures of program performance are 3 percent or less. Treated acreage and environmental points decline slightly, while overall net income support increases slightly. For larger minimums, the decline in participation slows while treated acreage and environmental points continue to decline slowly and net income support continues to increase slowly.

Sensitivity of Index Weights

Index weights determine the relative importance of various environmental problems and, because payments are based in part on environmental scores, can exert a strong influence on which conservation treatments are actu-ally undertaken. To get some sense of how much index weights influence outcomes, we test the sensitivity of the model to changes in index weights. For the sensitivity analysis, we place each index component (and all associated subcomponents) into one of four categories: soil erosion, water quality, grazing land health, and wildlife habitat (table A5. 1). For each category, we re-estimate the model doubling weights in one category while weights in other categories are reduced, in total, by an equal amount. Weights are reduced in proportion to the original weight for each subcomponent. The procedure ensures that the total number of possible points is unchanged and that the relationship among model subcomponents (ones for which weights are not doubled) is unchanged.

National results show that the effect of doubling index weights for component groups varies. For grazing land health, doubling the component weight would increase the overall number of acres treated and the overall number of environmental points earned (figure A5.2). More specifically, the number of acres treated for grazing land health would increase from about 125 million with the base weights to about 175 million with the grazing land health index weight doubled. This result follows from the fact that a large acreage is eligible for grazing land health treatment (about 220 million acres; see table A2. 1 in appendix 2), but many of those acres would be costly to treat (median WTA is about $50 per acre—the highest of any treatment; see table A3. 1 in appendix 3). So, higher payments that come with a higher score make a significant difference in the number of acres treated. The increase in acreage treated and environmental points earned for grazing land health are larger than decreases associated with other treatments (where index weights were reduced).

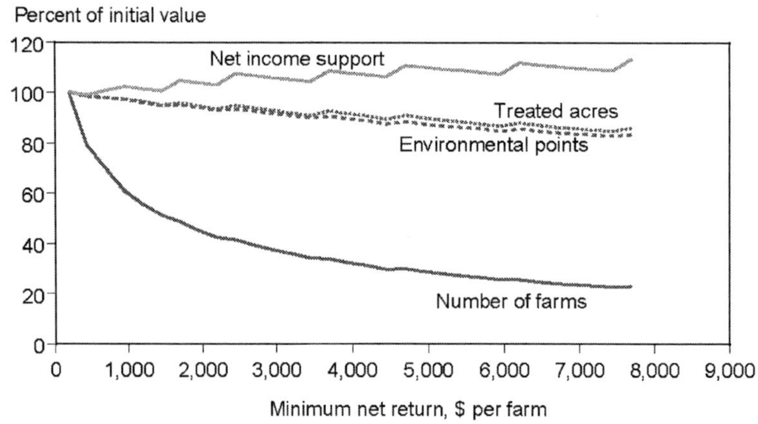

Source: USDA, Economic Research Service.

Figure A5.1. Effect of increasing minimum net return, improved performance scenario.

Table A5.1. Index component grouping for sensitivity analysis.

Group	Index components	Combined weight
Grazing land health	Grazing land health	.06
Soil erosion	Wind erosion (cropland) Water erosion (cropland)	.60
Water quality	Nutrient management (cropland) Pest management (cropland) Nutrient management and Riparian erosion (grazing land)	.54
Wildlife habitat	Wildlife habitat (all land)	.60

Source: USDA Economic Research Service.

Table A5.2. Sensitivity to change in index weights.

	Program payments	Number of farms	Net income support	Treated acres	Environmental points
	Million $	*1,000*	*Million $*	*Millions*	*Millions*
Base weights	14,960	1,132	9,088	411	19
Double grazing land health weight	14,858	1,090	8,563	413	23
Double soil erosion weight	15,082	1,104	8,548	375	12
Double water quality weight	14,930	1,173	9,643	370	27
Double wildlife habitat weight	15,066	996	11,350	296	13

Source: USDA Economic Research Service.

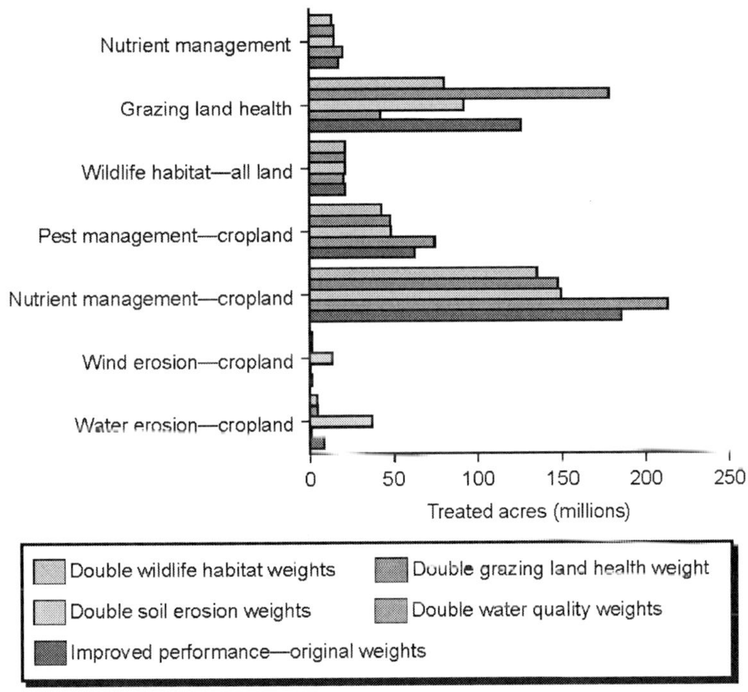

Source: USDA, Economic Research Service.

Figure A5.2. Treated acres, sensitivity to change in index weights.

In contrast, doubling the weight on the wildlife habitat index components would reduce overall acreage treated and the overall number of environmental points earned. Doubling the wildlife habitat weights would not result in additional treatment because only a relatively small number of acres are eligible for this treatment to begin with (about 20 million acres; see table A2.1) and nearly all of these acres are treated in the *Improved Performance* scenario with the base weights. Meanwhile, reducing other weights would lower payments to other treatments, reducing treated acreage and environmental points.

In terms of net income support, doubling the weight on the grazing land health component would result in a small overall reduction while doubling the wildlife habitat weight would result in an increase of more than $1 billion in net support. Net income support is increased when the wildlife habitat weight is increased because payments to producers who adopt wildlife-related practices increase even if additional acres can't be treated for wildlife-related concerns.

Beef producers gain additional support from increases in the weights for grazing land health and wildlife habitat but lose support when soil erosion or water quality is targeted for increase (figure A5.3). The situation is different for crop producers: Increases in weight for water quality components increases their income support while increases for grazing land health or wildlife habitat decrease their overall level of income support. Regions with large acreages of grazing land tend to receive greater income support when weights for grazing land health or wildlife habitat are increased (figure A5.4). Regions with large crop acreages tend to receive more income support when the weight given to water quality is increased.

Sensitivity of Environmental Hurdle Rates

Environmental hurdle rates in the *Good Performance* scenario define environmental performance that is good enough to qualify for some level of payment. Our initial *Good Performance* scenario assumes that hurdle rates are (1) devised on a region-by-region basis and (2) define the hurdle rate in a way that includes farms which account for roughly one-half of all agricultural land. Here, we consider two changes: (1) hurdle rates set for the Nation as a whole and (2) hurdle rates set for varying levels of inclusiveness.

Given a $15 billion budget, the number of participating farms is larger when hurdle rates are calculated at a regional, rather than national level (figure A5.5). Depending on the level of the hurdle rate (and assuming a $200 minimum per-farm return for participating farms), between 100,000 and 200,000 additional farms participate when hurdle rates are set regionally. As the hurdle rates are adjusted to increase the amount of land in "good actor" status, participation rises with both regionally and nationally defined hurdle rates. As the hurdle rates become increasingly inclusive, participation becomes attractive to a broader group of producers because more producers can participate without taking any additional conservation action or incurring additional cost. Treated acreage, on the other hand, declines rapidly as the hurdle rate becomes more inclusive (figure A5.6) and more producers can qualify for payments without undertaking additional conservation treatments.

As the budget is increasingly devoted to these good performance payments, less is available for leveraging additional conservation effort. To limit payments to $15 billion, the payment rate must be lower and the number of acres treated declines. Finally, environmental performance, as measured in index points, is largest when hurdle rates are set to include about 30 percent of land but also declines when the environmental hurdle rate is more inclusive (figure A5.7).

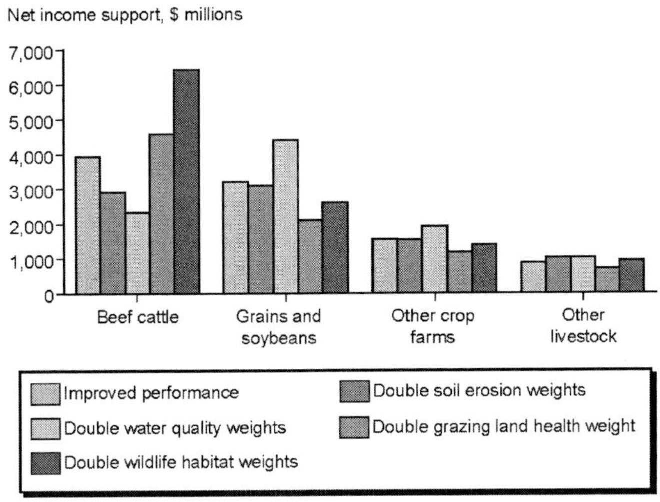

Source: USDA, Economic Research Service.

Figure A5.3. Net income support by scenario and primary commodity—sensitivity to change in index weights.

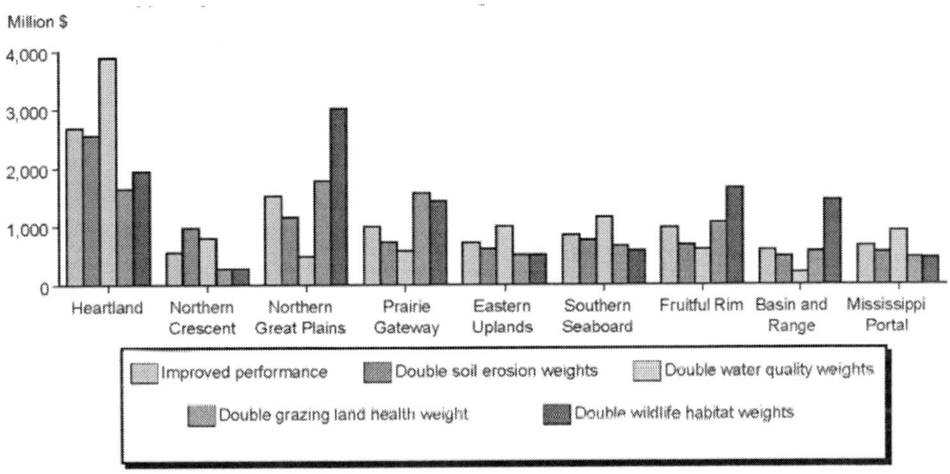

Source: USDA, Economic Research Service.

Figure A5.4. Net income support by scenario and ERS resource region.

Finally, income support rises as the hurdle become increasingly inclusive (figure A5.8). Payments for good performance are the key engine of growth for income support. Even as good performance payments rise, however, reduced conservation treatment activity means that income support associated with the "profit" from conservation treatment declines. As a result, net income support rise less than do good performance payments.

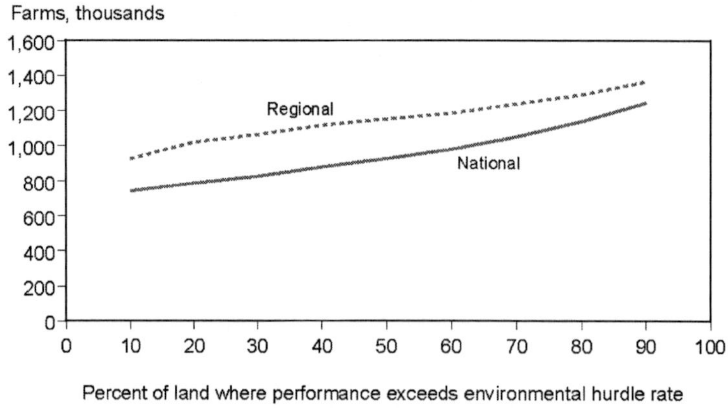

Source: USDA, Economic Research Service.

Figure A5.5. Particpating farms—national and regional hurdle rates.

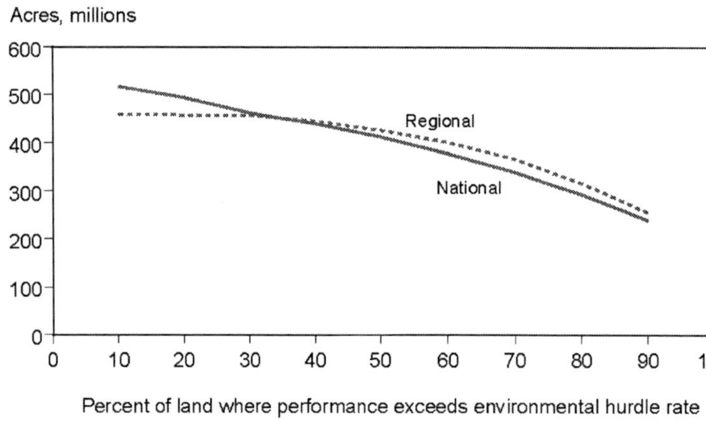

Source: USDA, Economic Research Service.

Figure A5.6. Acres treated—national and regional hurdle rates.

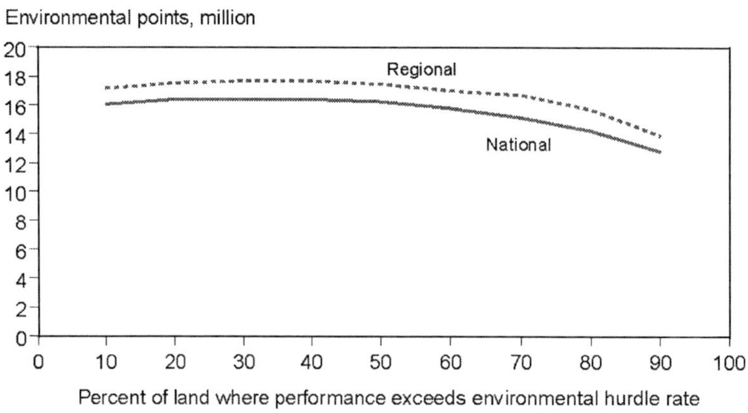

Source: USDA, Economic Research Service.

Figure A5.7. Environmental points—national and regional hurdle rates.

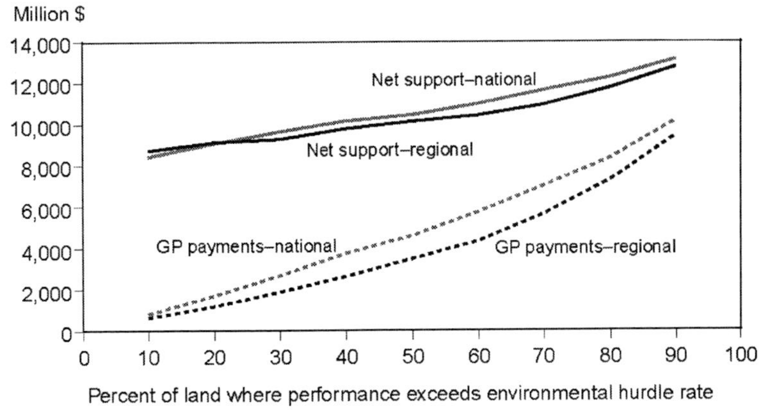

Source: USDA, Economic Research Service.

Figure A5.8. Net income support and good performance payments— national and regional hurdle rates.

REFERENCES

Alexander, R. B., Smith, R.A. & Schwartz, G. E. (2000). "Effect of stream channel size on the delivery of nitrogen to the Gulf of Mexico," *Nature, 403*, 758-761.

Atwood, J., Buland, D., Cattaneo, A., Esser, A., Knight, L., Toomer, L., Vik, D. & Whitcomb, J. (2003). *Environmental Quality Incentives Enhancement Program: Benefit-Cost Assessment.* U.S. Department of Agriculture, Natural Resources Conservation Service, May.

Atwood, J., Claassen, R., Featherstone, J., Knight, L., Moore, L. & O'Neill, G. (2004). *Conservation Security Program: Benefit-Cost Assessment.* U.S. Department of Agriculture, Natural Resources Conservation Service, May.

Claassen, R., Breneman, V., Bucholtz, S., Cattaneo, A., Johansson, R. & Morehart, M. (2004). *Environmental Compliance in Agricultural Policy: Past Performance and Future Potential.* U.S. Department of Agriculture, Economic Research Service, AER-832.

Claassen, R., Hansen, L., Peters, M., Breneman, V., Weinberg, M., Cattaneo, A., Feather, P., Gadsby, D., Hellerstein, D., Hopkins, J., Johnson, P., Morehart, M. & Smith, M. (2001). *Agri-Environmental Policy at the Crossroads.* U.S. Department of Agriculture, Economic Research Service, AER-794.

Feather, P. & Hellerstein, D. (1997). "Calibrating Benefit Function Transfer to Assess the Conservation Reserve Program," *American Journal of Agricultural Economics, 79(1)*, 151-162.

Feather, P., Hellerstein, D. & Hansen, L. (1999). *Economic Valuation of Environmental Benefits and the Targeting of Conservation Programs: The Case of the CRP.* U.S. Department of Agriculture, Economic Research Service, AER-778.

Gloss, Don, W., Robert, L. Kellogg, Joaquin Sanabria, Susan Wallace, & Walt Kniesel. (1998). *The National Pesticide Loss Database: A Tool for Management of Large Watersheds.* Poster Presentation at the 53rd Annual SWCS Conference San Diego, California, July 5-9, 1998. Data available at http://www.nrcs.usda.gov/ technical/ land/pubs/gosstext.html.

Hansen, L., Breneman, V., Davison, C. & Dicken, C. (2002). "The Cost of Soil Erosion to Downstream Navigation." *Journal of Soil and Water Conservation, 57(4)*, 205-212.

Piper, Steven, & Paul, C. Huszar. (1989). "Re-examination of the offsite costs of wind erosion in New Mexico." *Journal of Soil and Water Conservation.* July-August, 332-334.

Ribaudo, M. O., Colacicco, D., Langner, L., Piper, S. & Schaible, G. (1990). *Natural Resources and Natural Resource Users Benefit from the Conservation Reserve Program.* U.S. Department of Agriculture, Economic Research Service, AER-627.

Smith, R., Schwarz, G. & Alexander, R. (1997). "Regional interpretation of water-quality monitoring data," *Water Resources Research, 33(12)*, 2781-2798.

U.S. Department of Agriculture. (2006). *2007 Farm Bill Theme Papers: Conservation and the Environment.* Available at: www.usda.gov/documents/FarmBill07consenv.pdf.

Williams, J. R., Putman, J. W. & Dyke, P. T. (1985). "Assessing the Effect of Soil Erosion on Productivity with EPIC." In *Erosion and Soil Productivity*, ASAE Publication 8-8 5, American Society of Agricultural Engineers, St. Joseph, MO.

End Notes

[1] Commercial farms are large family farms with sales above $250,000 per year and some nonfamily farms organized as cooperatives or nonfamily corporations. Intermediate farms have annual sales below $250,000 and the operator reports farming as his or her major occupation. Rural residence farms have annual sales below $250,000 where farming is considered a secondary activity both in terms of resources invested in the farm and the amount of income it contributes to the farm household.

[2] The hurdle rate could have been calculated at other geographic scales. For a comparison of basic results using regional and national hurdle rates, see Appendix 5: Sensitivity Analysis.

[3] Note that grazing land cannot be enrolled and that wildlife habitat resource concerns cannot be addressed through compliance scenarios.

[4] Producers have multiple options in each of the scenarios except *Extended Compliance.*

[5] While "T" is generally associated with soil productivity, we use it as a more general threshold for treatment. That is due to practical considerations; treatment for soil erosion concerns has typically meant treatment to the T level in USDA programs from which cost data are drawn.

[6] NRCS refers to these as resource concerns. Because we have used the term "resource concern" to refer to the broader concerns of "water quality" or "air quality," we refer to these as NRCS classification codes to avoid confusion.

[7] In the text WTA is ofen refered to as economic cost.

In: U.S. Farms and Conservation Programs
Editor: Lucas A. Savarese

ISBN: 978-1-61209-927-9
© 2011 Nova Science Publishers, Inc.

Chapter 2

PARTICIPATION IN CONSERVATION PROGRAMS BY TARGETED FARMERS: BEGINNING, LIMITED-RESOURCE, AND SOCIALLY DISADVANTAGED OPERATORS' ENROLLMENT TRENDS

Cynthia Nickerson and Michael Hand

ABSTRACT

Beginning, limited-resource, and socially disadvantaged farmers make up as much as 40 percent of all U.S. farms. Some Federal conservation programs contain provisions that encourage participation by such "targeted" farmers and the 2008 Farm Act furthered these efforts. This chapter compares the natural resource characteristics, resource issues, and conservation treatment costs on farms operated by targeted farmers with those of other participants in the largest U.S. working-lands and land retirement conservation programs. Some evidence shows that targeted farmers tend to operate more environmentally sensitive land than other farmers, have different conservation priorities, and receive different levels of payments. Data limitations preclude a definitive analysis of whether efforts to improve participation by targeted farmers hinders or enhances the conservation programs' ability to deliver environmental benefits cost effectively. But the different conservation priorities among types of farmers suggest that if a significantly larger proportion of targeted farmers participates in these programs, the programs' economic and environmental outcomes could change.

Keywords: Conservation programs, Environmental Quality Incentives Program (EQIP), Conservation Reserve Program (CRP), Wetlands Reserve Program (WRP), beginning farmers, limited-resource producers, socially disadvantaged farmers and ranchers

ACKNOWLEDGMENTS

The authors thank David Nulph and Vince Breneman (both from USDA's Economic Research Service) for assistance with analyses using geographic information systems data. We also thank the following for their helpful comments on earlier drafts: Mary Ahearn, David McGranahan, and Marca Weinberg, all from ERS; Alex Barbarika from USDA's Farm Service Agency; Cathy Kling from Iowa State University; Dayton Lambert from the University of Tennessee; and Felix Spinelli from USDA's Natural Resources Conservation Service. Angela Anderson (editorial) and Victor B. Phillips, Jr. (layout and design) provided production assistance.

SUMMARY

Targeted farmers include those with 10 or fewer years of experience, farmers with limited farm sales and income, and farmers belonging to segments of the population that have historically been subject to discrimination, such as African American, American Indian, Alaskan Native, Hispanic, Asian American, or Pacific Island farmers. Targeted farmers make up as much as 40 percent of all U.S. farms, and, although many participate in conservation programs, targeted farmers typically have not participated in government agricultural programs at the same rate as other farmers. To help offset potential barriers to participation, USDA offers targeted farmers more favorable payment and enrollment terms in conservation programs than are available to other farmers.

What is the Issue?

Farm legislation in both 2002 and 2008 encouraged targeted farmers to participate in conservation programs by making them eligible for more favorable payment and enrollment terms than other farmers received. Such Federal provisions can alter program outcomes in unintended ways if targeted farmers adopt different conservation practices, address different environmental needs, or operate land that is more or less environmentally sensitive than the land operated by other farmers. Targeting certain farmer types could result in tradeoffs between environmental performance, cost-effective delivery of program benefits, and improved access to Federal conservation programs. This chapter addresses this issue by examining participation patterns in the Environmental Quality Incentives Program (EQIP), Conservation Reserve Program (CRP), and Wetlands Reserve Program (WRP) to provide information on the types of farmers who enroll, the geographic distribution of participants, the types and costs of conservation practices they implement, the resource issues they address, the natural resource characteristics associated with their farms, and whether different types of farmers participate in different ways. Participation rates are measured where possible based on both the number of farms and acres enrolled in conservation programs, as these two measures can provide very different pictures of targeted farmer participation. Those three conservation programs account for 74 percent of authorized conservation spending in the 2008 Farm Act.

What Did the Study Find?

During 2004-07, targeted farmers participated differently in conservation programs than did other farmers. While not definitive, evidence shows that targeted farmers tended to operate more environmentally sensitive land than other farmers, had different conservation priorities, and received different levels of payments. Those differences suggest that economic and environmental outcomes could change if the proportion of targeted farmers enrolled in the programs increases significantly.

Enrollment Patterns

Beginning and limited-resource farmers—two of the three targeted-farmer groups—were less likely to participate in EQIP than was the U.S. farm population as a whole (data on socially disadvantaged farmer participation in EQIP were not available). These two farmer types operated about 27 percent of all farms in 2006 but held 12 percent of EQIP contracts. This participation trend was observed in nearly every region of the country, suggesting that policies that make it easier for these farmers to enroll could increase participation. One new policy aimed at encouraging participation, however, is likely to have little effect. The 2008 Farm Act requires that 5 percent of EQIP funds be set aside annually for beginning farmers, but beginning farmers have typically received more than 10 percent of EQIP payments annually in recent years.

Like their participation patterns in EQIP, targeted farmers enrolled disproportionately fewer farms in conservation programs that retire land from production. Twenty-two percent of farms operated by all three groups of targeted farmers were enrolled in the CRP and WRP, even though they operate 31 percent of all farms. A different pattern emerges, however, when the amount of enrolled acreage is evaluated instead of the number of enrolled farmers: Targeted farmers enrolled disproportionately more acreage in CRP and WRP than other farmers. Targeted farmers operated 15 percent of farmland acres in 2007 but controlled 17 percent of acres enrolled in these programs.

Environmental Problems and Priorities for Treatment

Beginning farmers in the Delta region (the only area for which we could analyze soil data) who participated in EQIP tended to enroll more highly erodible land than other participating farmers. This trend suggests that conservation efforts by these farmers could provide more program benefits than efforts by other farmers. The available data, however, make it difficult to determine with certainty if targeting these farmers would increase program benefits because the characteristics of program participants may not represent this farmer group as a whole. If, in general, few beginning farmers operate highly erodible land (and the few that do have "self-selected" and have already chosen to enroll), targeting more of these farmers for enrollment may not provide more conservation benefits than are provided by other farmer types.

Conservation priorities of farmers participating in conservation programs differ by farmer type. Beginning and limited-resource farmers enrolled in EQIP addressed livestock forage and health needs and plant productivity/ quality issues more often than did other farmer types. And although both beginning and limited-resource farmers participating in EQIP were more likely to farm closer to quality-impaired waters, the limited-resource farmers were less likely to address water quality problems. Several possible reasons may explain these tendencies. First, limited-resource farmers may face financial or other constraints in adopting practices that might improve water quality. Second, these farmers may derive more direct benefits by focusing on other issues (like improving plant health and vigor). Finally, limited- resource farmers tend to operate smaller farms that are not subject to the same regulatory requirements facing larger farm operators who use EQIP to fund water pollution reduction practices (e.g., pollutant discharge regulations for concentrated animal feeding operations). In land retirement programs, targeted farmers of all types were more likely than other farmer types to be located in areas where proposed conservation efforts were expected to achieve the greatest reduction in soil erosion and the greatest improvement in water quality.

Costs of Treatment

The size of conservation payments varied among farmer groups.

- In EQIP, average payments (which represent the cost to government) to beginning farmers were significantly higher than the average payments to other farmers, while payments to limited-resource farmers were significantly lower. Both beginning and limited-resource farmers implemented a larger number of conservation practices than other farmers did, but the scale of those practices tended to be smaller.
- In CRP and WRP, targeted farmers enrolled a greater share of operated acreage and received smaller per acre payments, but the number and types of conservation practices adopted did not differ significantly from those of other farmers.

How Was the Study Conducted?

The analysis relied on USDA data from EQIP and CRP administrative records, the 2007 Census of Agriculture, and the National Resources Inventory (NRI). The analysis also used data from the Agricultural Resource Management Survey (ARMS), which is conducted annually by the National Agricultural Statistics Service and the Economic Research Service.

The EQIP and CRP contract data identified the resource problems that received treatment and, for EQIP, how payments and adoption of practices varied between beginning/limited-resource producers and other participant types. The analysis used the ARMS data from 2004-07 to analyze payment and acreage enrollments in CRP and WRP by farmer type and to summarize the characteristics of targeted farmers. The 2001 ARMS data on conservation practice adoption (the most recent year data were available) were used to examine conservation practice adoption patterns in land retirement programs. The census, NRI, and other data characterized the distribution of farmer types relative to measures of environmental conditions.

EQIP, CRP, ARMS, NRI, and census data were used to characterize differences among current conservation program participants and to suggest that targeted and other farmers may differ in their ability to provide environmental benefits cost effectively. Providing firm answers about the impacts of favoring particular farmer types would require more information, including quantitative estimates of the environmental benefits provided by different farmer types and whether targeted participants are more cost-effective providers of benefits than nonparticipants. Also, targeted farmers' acreages are disproportionately small, and information about program participants' farm sizes would be needed to distinguish whether differences between farmer types are due to the type of farmer or farm size.

INTRODUCTION

The Federal Government has a long history of providing assistance to farmers and ranchers to encourage the adoption of practices that reduce unintended negative environmental spillovers from agricultural production. Much of this conservation assistance is provided through voluntary programs that help finance the installation of conservation structures (e.g., riparian buffers and grassed waterways) and the adoption of environmentally friendly land management practices or that provide technical support to identify conservation needs and develop implementation plans. The 2008 Food, Conservation, and Energy Act (2008 Farm Act) is projected to increase conservation program funding by more than 17 percent. If appropriated at authorized levels, $11.5 billion in conservation assistance will go to working farm and ranchland that remain in production, and $13 billion will go to land retirement programs.

Both the 2002 and 2008 Farm Act legislation included provisions that ensure accessibility for all eligible farmers and ranchers to conservation programs. These provisions provide favorable payment terms to reduce unintentional barriers to participation for farmers who may face unique circumstances, such as limited farming experience, limited financial resources, and limited opportunities. The favorable payment terms are available to beginning, limited-resource, or socially disadvantaged segments of the farm population ("targeted farmers" for the purposes of this chapter).

Targeted farmers' characteristics can affect their participation in conservation programs (see box, "Defining Targeted Farmers"). For example, recent Economic Research Service (ERS) research shows that beginning farmers tend to have smaller farms, lower levels of onfarm income, and different personal and household characteristics than do established farms (Ahearn and Newton, 2009) and that these differences can affect decisions about conservation activities (Lambert et al., 2006; Caswell et al., 2001).

Providing favorable payment terms to particular types of farmers may make Federal conservation programs more accessible if targeted farmers could not participate without them. Some evidence suggests that encouraging participation by targeted farmers may also alter the environmental benefits generated by conservation programs. Comparing the distribution of farmer types and highly erodible land suggests that socially disadvantaged farmers are more likely than other farmers to be located in counties where a higher proportion of cropland is highly erodible (figure 1).[1] When farmers are more likely to operate marginal land, targeting them for participation could increase the environmental benefits provided by conservation programs.

Changing program participation rates among targeted farmers could also affect program costs, and whether costs increase or decrease depends on the cost effectiveness of practices implemented by new participants. If the farmers who enroll in conservation programs provide environmental benefits at the lowest cost, encouraging their participation achieves both economic and accessibility goals. If applicants from targeted groups provide benefits at a higher cost, encouraging their participation may improve access at the expense of economic goals.

DEFINING TARGETED FARMERS

USDA has established definitions for beginning, limited-resource, and socially disadvantaged farmers (referred to as "targeted" for the purposes of this chapter). Our definitions correspond to USDA definitions to the extent that available data allow us to do so.

Beginning farmers and ranchers (BF)—Using data from USDA's Agricultural Resource Management Survey (ARMS) (see appendix, "Data Sources"), we can identify up to three operators per farm and define beginning farmers and ranchers as operators with not more than 10 years of experience. In 2006, however, ARMS identifies more than one operator in only one version of the survey. For that year, we define beginning farms as those operated by a primary operator with not more than 10 years of experience. USDA generally defines beginning farmers and ranchers as those who materially and substantially participate in farm or ranch operations, but have not operated a farm or ranch for more than 10 consecutive years. If operated by an entity, this requirement applies to all members of the entity (USDA, 2007).

Limited-resource farmers and ranchers (LR)—Farmers and ranchers are defined as limited resource if in each of the last 2 years they earned less than $105,000 in gross farm sales (adjusted for infl ation) and had low household income (USDA, 2007; Hoppe et al., 2007). Low household income means that the household income was less than the national poverty level for a family of four or was less than half the county median household income in the 2 previous years.

Socially disadvantaged farmers and ranchers (SDA)—In this chapter, we define a farm as operated by a socially disadvantaged farmer and rancher if the principal operator is African American, American Indian, Alaskan Native, Hispanic, Asian American, or Pacific Islander. We do not include women in this definition. While some socially disadvantaged farmer definitions include groups subject to gender prejudice, our definition is consistent with the Conservation Title in the 2008 Farm Act, which excludes women (unless they meet the socially disadvantaged definition some other way). Prior to the 2008 Farm Act, conservation programs administered by USDA did not use socially disadvantaged farmer definitions. USDA does, however, have various credit, insurance, and outreach programs to improve access to USDA programs for such farmers and ranchers (Dismukes et al., 1997a).

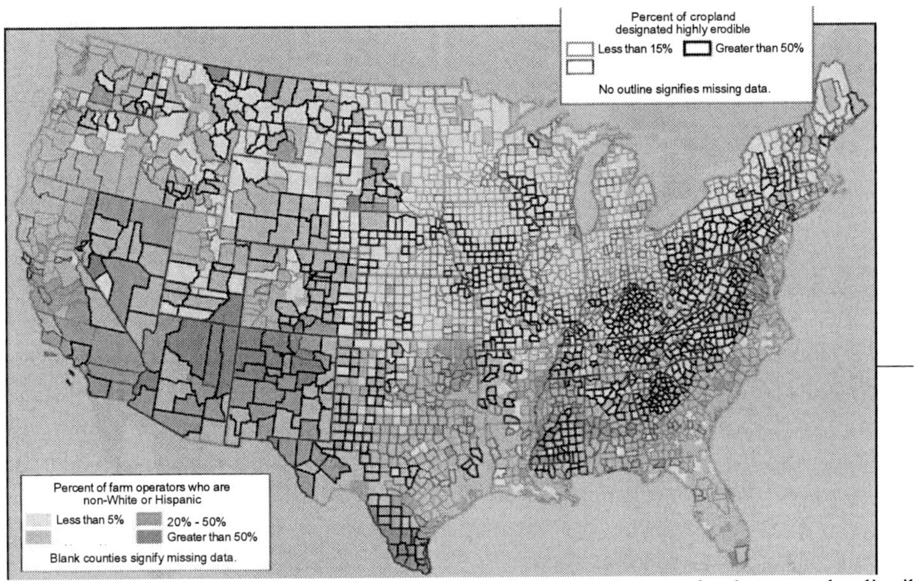

Note: Counties in dark green with a black outline identify the greatest overlap between the distribution of selected socially disadvantaged groups and highly erodible cropland. Although some rangeland may also be subject to erosion, erodibility data are available only for cropland.

Source: 2007 Census of Agriculture (non-White and Hispanic operators) data and 1997 National Resources Inventory (highly erodible cropland) data.

Figure 1. Distribution of selected farmer types and highly erodible cropland, by county.

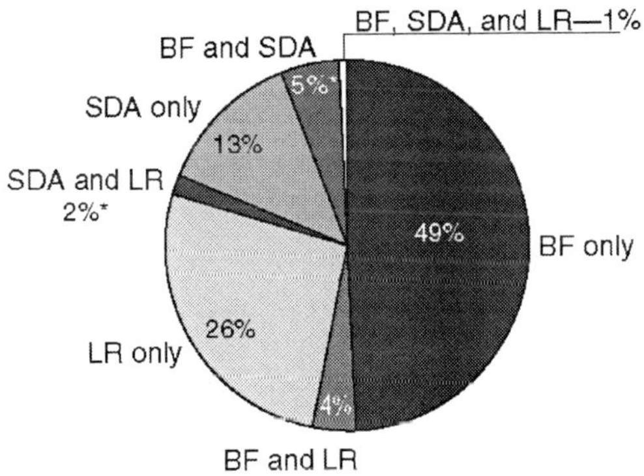

* Coefficient of variation is between 25 percent and 50 percent.

BF=Beginning farmers; SDA=Socially disadvantaged farmers; LR=Limited-resource farmers.

Note: Farms operated by women are not included with socially disadvantaged farmers, unless they otherwise meet the SDA definition.

Source: ERS calculation based on USDA's pooled 2005-07 Agricultural Resource Management Surveys, Phase III, conducted by the National Agricultural Statistics Service and the Economic Research Service.

Figure 2. Farms operated by targeted farmers and ranchers.

* Coefficient of variation is between 25 percent and 50 percent.

BF=Beginning farmers; LR=Limited-resource farmers; SDA=Socially disadvantaged farmers.

Notes: Farms operated by women include farms with female primary operators that do not meet the definitions of targeted farmers. The "all other farms" category includes farms where women are not primary operators. Totals may not sum to 100 due to rounding.

Source: ERS calculations based on USDA's 2007 Agricultural Resource Management Survey, Phase III, conducted by the National Agricultural Statistics Service and the Economic Research Service.

Figure 3. Share of total farms and production, by operator status.

This chapter examines participation patterns of targeted farmers relative to other farmers in major Federal conservation programs. We focus on the extent of their participation, the geographic distribution of participants, the types and costs of conservation practices implemented, and resource issues farmers typically face and address. Although the available data preclude a definitive assessment of economic tradeoffs from improving accessibility, this chapter improves our understanding of how these farmers use and impact conservation programs.

What We Know about Targeted Farmers

Targeted farmers include beginning, limited-resource, and socially disadvantaged farmers, and most farm operators identify with just one of these three groups (figure 2). During 2005-07, the largest proportion (59 percent) of farms operated by targeted farmers was beginning-farmer operations. Socially disadvantaged (SDA) farmers accounted for 21 percent and limited-resource farmers, 33 percent. These percentages sum to more than 100, signaling that some farmers belong to multiple groups. For example, 5 percent of these farmers were both beginning and socially disadvantaged farmers, and 4 percent were both beginning and limited-resource farmers. Only 1 percent of these farms were operated by farmers who belonged to all three groups.

The characteristics of these operations can improve our understanding about why targeted farmers might participate in conservation programs differently than other farmers. Targeted farmers operate a sizable proportion of land. In 2007, they operated about 677,000 farms—

nearly 31 percent of the Nation's 2.1 million farms (figure 3). Yet, these farmers account for a disproportionately small amount of farm production value, contributing only 12 percent of the nearly $290 billion in farm commodities produced by all U.S. farms in 2007. This smaller farm production value refl ects the fact that a larger proportion of these farmers operate small-scale family farms (defined as those with less than $250,000 in sales) and that a smaller proportion of beginning and socially disadvantaged farmers identify farming as their major occupation—residential or lifestyle farms—compared with other farmer types (table 1).[2] Limited-resource farmers are more likely to report farming as their major occupation; however, fewer farms are operated by them. Low-sales farms and farmers who devote time to nonfarming occupations may face financial or labor constraints that preclude use of conservation practices that are capital- or management-intensive.

Although their farms are typically small, some targeted farmers operate relatively large farms. In 2007, the median acres operated—the midpoint of the range of acres operated—by targeted farmers ranged from 41 to 75 acres, compared with 94 acres for small family farms and 118 acres for farms of all sizes operated by other farmer types (table 2). The high average number of acres operated, relative to median number of acres operated, suggests that all farmer types include relatively large farms. In particular, a few socially disadvantaged farms are organized as nonfamily farms that operate larger farms, on average. Also, socially disadvantaged farms specializing in livestock are about 150 acres larger, on average, than socially disadvantaged farms in general.

Beginning and limited-resource farmers are somewhat more likely than other farmers to report no production value. The share of these farmers reporting zero production value varies between 26 and 28 percent, compared with 23 percent of small family farms operated by other farmer types.[3,4] Recent research that analyzes production and conservation behavior revealed that decisions not to produce crops or livestock are positively associated with conservation payments (Lambert et al., 2006). This suggests that conditions that temporarily rule out farming may not preclude conservation program participation, although a lack of production over several years could make some farmers ineligible to participate.

For farms reporting a positive production value, commodity production by targeted farmers appears similar to small family farms, at least at first glance. Livestock accounts for most of the production value for farmers of all types, and more than half of socially disadvantaged farmers specialize in livestock (table 2). Specialization in beef cattle and other grazing livestock is common among small farms, due to the low labor and low cost requirements (Hoppe et al., 2007). Of farmers with a majority of their production value from crop sales, a larger proportion of socially disadvantaged farmers focus on highvalue specialty crops, such as vegetables, fruits and tree nuts, and nursery and greenhouse products.

Various measures of financial performance reveal that not all targeted farmers are alike, and some are more likely to face issues that prevent them from adopting conservation practices. Limited-resource farmers may be more likely to face liquidity constraints due to lower net farm income and a higher operating expense ratio compared with other farmer types (table 2). Limited- resource farmers rely most heavily on off-farm income, the majority of which comes from unearned sources (Social Security, pensions, dividends, interest, and rent) (Hoppe et al., 2007). Most socially disadvantaged and beginning farmers operate residential farms, and most of their off-farm income comes from earned wages or self-employment (Hoppe et al., 2007).

Table 1. Number of farms and production value, by farm typology and farmer status, 2007.

Item	Small family farms[1]				Large family farms	Nonfamily farms	All farms
	Retirement	Residential and lifestyle	Farming occupation	Total			
All farms:							
Total farms (number)	408,306	1,004,679	535,779	1,948,764	205,368	42,633#	2,196,766
Percent of all farms	18.6	45.7	24.4	88.7	9.4	1.9	100
Percent of small family farms	21.0	51.6	27.5	100	—	—	—
Value of production (percent of U.S. total)	1.5	4.2	11.1	16.8	67.5	15.7	100
All BF, LR, and SDA farmers:[2]							
Total farms (number)	166,692	335,229	150,088	652,009	16,993	8,002 #	677,005
Percent of all BF, LR, and SDA farmers	24.6	49.5	22.2	96.3	2.5	1.2	100
Percent of small family farms	25.6	51.4	23	100	—	—	—
Value of production (percent of U.S. total)	0.4	0.9	1.6	2.9	6.8	2.7	12.4
Beginning farmers:							
Total farms (number)	47,582	254,190	54,089	355,861	13,715	7,252 A	376,829
Percent of all beginning farmers	12.6	67.5	14.4	94.4	3.6	1.9	100
Percent of small family farms	13.4	71.4	15.2	100	—	—	—
Value of production (percent of U.S. total)	0.1	0.7	0.7	1.5	4.8	1.6	7.8
Limited-resource farmers:							
Total farms (number)	120,333	49,328	85,331	254,992	0	0	254,992
Percent of all limited-resource farmers	47.2	19.3	33.5	100	0	0	100
Percent of small family farms	47.2	19.3	33.5	100	—	—	—
Value of production (percent of U.S. total)	0.2	0.1	0.8	1.1	0	0	1.1
Socially disadvantaged farmers:							

Table 1. (Continued)

Item	Small family farms[1]				Large family farms	Nonfamily farms	All farms
	Retirement	Residential and lifestyle	Farming occupation	Total			
Total farms (number)	16,842	72,388	24,677	113,907	4,129	818 #	118,854
Percent of all SDA farmers	14.2	60.9	20.8	95.8	3.5	0.7	100.0
Percent of small family farms	14.8	63.6	21.7	100	—	—	—
Value of production (percent of U.S. total)	0.1	0.2	0.3	0.6	2.5	1.2	4.2
Socially disadvantaged farmers, including women in the definition:							
Total farms (number)	81,048	183,439	76,965	341,452	8,992	3,940 ##	354,383
Percent of all SDA farmers	22.9	51.8	21.7	96.4	2.5	1.1	100
Percent of small family farms	23.7	53.7	22.5	100	—	—	—
Value of production (percent of U.S. total)	0.2	0.4	0.6	1.1	4.1	1.6	6.8
Other farmer types: [3]							
Total farms (number)	241.614	669,450	385,691	1,296,755	188,375	34,631 #	1,519,761
Percent of all other farmer types	15.9	44.0	25.4	85.3	12.4	2.3	100
Percent of small family farms	18.6	51.6	29.7	100	—	—	—
Value of production (percent of U.S. total)	1.1	3.3	9.5	13.9	60.7	13.0	87.6

Coefficient of variation (CV=Standard Error / Estimate) is between 25 percent and 50 percent; ## CV is between 50 percent and 75 percent.

— Data not applicable. BF=Beginning farmers; LR=Limited-resource farmers; SDA=Socially disadvantaged farmers.

Note: Totals may not sum to 100 due to rounding.

[1] Small family farms include all farms with less than $250,000 in gross sales.

[2] Includes all farms operated by beginning, limited resource, or socially disadvantaged farmers. Does not include farms operated by women who do not meet other definitions of socially disadvantaged farmers. Some farmers may fit the definition of more than one category and are included in the counts for each.

[3] Includes all farms that do not meet the definition of beginning, limited-resource, or socially disadvantaged farmers. Includes farms operated by women who do not meet other definitions of socially disadvantaged farmers.

Source: ERS calculations based on USDA's 2007 Agricultural Resource Management Survey, Phase III, version 1, conducted by the National Agricultural Statistics Service and the Economic Research Service.

Table 2. Selected farm characteristics and financial indicators, by farmer status, 2007.

Item	Selected farmer types[1]				Other farmer types[2]	
	Limited resource	Socially disadvantaged	Socially disadvantaged, including women	Beginning farmer	Small family farms	All farms
Acres operated:						
Mean (average)	153	314	234	163	270	509
Median	75	48	49	41	94	118
Percent of U.S. farmland	4.3	4.5	9.5	6.8	38.6	85.2
Percent of U.S. cropland	3.4	2.1	5.3	5.6	34.1	89.2
			Percent of group			
Tenure:						
Full owner	78.7	68.7	78.6	77.6	65.8	59.9
Part owner	14.4	20.9	16.3	13.1	29.7	34.2
Tenant	D	10.4#	5.1#	9.3	4.5	5.9
Commodity specialization:[3]						
Livestock	47.5	53.2	44.4	43.9	45.4	44.5
Crop	24.4	26.1	20.4	26.6	29.8	33.8
Type of crop—						
Field crops	18.8	12.1#	10.8	21.0	23.8	27.3
High value crops[4]	D	14.0	9.6	5.6	6.0	6.5
Whole farm CRP/WRP[5]	D	D	D	D	1.7	1.5
Zero value of production	25.9	20.4	33.2	28.1	23.2	20.2
Sales class:[6]						
$0 sales	17.6	17.6	23.6	20.4	15.8	13.6
$1 to $49,999	77.4	70.4	67.0	69.3	65.5	56.7
$50,000 to $99,999	D	4.6#	3.8	3.4	8.9	8.0
$100,000 to $174,999	D	1.9#	1.7	1.8	5.9	5.1
$175,000 to $499,999	0.0	3.0	2.2	2.9	4.0	9.5
$500,000 or more	0.0	2.6	1.0	2.3	0.0	7.1

Table 2. (Continued)

Item	Selected farmer types[1]				Other farmer types[2]	
	Limited resource	Socially disadvantaged	Socially disadvantaged, including women	Beginning farmer	Small family farms	All farms
Financial indicators:						
Net farm business income (average $)	2,448#	19,831#	10,810#	9,975	9,992	45,004
Net farm business income, less conservation payments (average $)	1 ,972#	1 9,497#	1 0,070#	9,475	9,089	43,842
Median net farm business income	2,409#	D	911##	673#	4,006	5,634
Operating expense ratio[7]	1.25	0.82	0.90	0.90	0.93	0.76
Debt to asset ratio[8]	0.03	0.09	0.07	0.12	0.04	0.08
Percent of household income from off-farm sources[9]	182.2	91.4	101.0	103.2	102.7	81.6

[#] Coefficient of variation (CV=Standard Error / Estimate) is between 25 percent and 50 percent; ##CV is between 50 percent and 75 percent. CRP=Conservation Reserve Program; WRP= Wetlands Reserve Program.

D = Not reportable due to small sample size or estimate validity concerns. 1The farmer categories are not mutually exclusive (figure 1).

[2] Includes all farms that do not meet the definition of beginning, limited resource, or socially disadvantaged farmers. Includes farms operated by women who do not meet other definitions of socially disadvantaged farmers.

[3] Commodity that accounts for at least half of the farm's value of production.

[4] Vegetables, fruits and tree nuts, and nursery and greenhouse.

[5] Defined as farms with all acres operated enrolled in CRP or WRP and no production.

[6] Sales classes are defined by gross farm sales, which include crop and livestock sales (regardless of who has claim on it), plus shares of production and government payments. For more detail, see Hoppe et al. (2007, p. 40).

[7] Total cash operating expenses divided by total gross cash farm income.

[8] Total liabilities divided by total assets.

[9] Calculated as (total off-farm household income divided by total household income)*1 00.

Source: ERS calculations based on USDA's 2007 Agricultural Resource Management Survey, Phase III, conducted by the National Agricultural Statistics Service and the Economic Research Service.

Targeted farmers are located across the United States, but some geographical concentration exists. The proportion of farms owned by beginning farmers ranges from 4 percent to 50 percent across U.S. counties, with higher proportions concentrated in Texas, Arizona, Colorado, and Florida (Ahearn and Newton, 2009). Socially disadvantaged farmers appear dispersed as a whole, but a large majority of African American farms are found in the South, and most American Indian farmers are found west of the Mississippi River. Farms operated by Asian/Pacific Islander operators are concentrated in California and Hawaii, while Hispanic-operated farms are concentrated in California, Colorado, Florida, New Mexico, and Texas (Dismukes et al., 1997a, 1997b).

CONSERVATION PROGRAMS AVAILABLE TO TARGETED FARMERS

All farmers have the opportunity to participate in Federal conservation programs, as long as they meet program eligibility requirements. Certain program provisions are specifically designed to increase access to working- lands programs for targeted farmers and ranchers. Understanding the differences in program incentives, eligibility, and purpose help explain observed patterns of program participation by members of these groups.

Federal agricultural conservation programs use a voluntary approach to improve environmental conditions related to agricultural production. Federal conservation funding is divided among land retirement, working-lands conservation, land preservation, and technical assistance programs. Of these, the Conservation Reserve Program (CRP), Environmental Quality Incentives Program (EQIP), and Wetlands Reserve Program (WRP) are the three largest programs and constitute about 70 percent of USDA conservation spending (see box, "Major Federal Conservation Programs").

Conservation programs use different approaches to address environmental concerns. Some goals of the major Federal conservation programs are similar—including reducing soil erosion, improving water and air quality, and enhancing wildlife habitat—but the approaches used to achieve these goals are different. The CRP and WRP seek to improve environmental quality by retiring environmentally sensitive land from production. EQIP helps producers address environmental concerns by providing financial assistance to install conservation structures and adopt management practices on agricultural lands that remain in production.

Some farmers may be more likely than others to participate in land retirement than in working lands programs, although many farmers may participate in both. The CRP and WRP have historically been popular, in part, because of the steady source of income that producers can receive from annual rental payments. Retired farmers and operators whose primary occupation is not farming may be more likely than other farmers to retire land and enroll in CRP and more likely to use land retirement conservation practices than working-lands practices (Lambert et al., 2006). As a working-lands program, EQIP may be most attractive to producers who want to improve environmental performance without reducing the operation's productive capacity, who operate land that is not eligible for CRP or WRP enrollment, who apply to CRP or WRP but are not accepted, or who favor shorter term conservation program contracts.

EQIP and land retirement programs have different enrollment methods and criteria. In CRP, highly erodible land is eligible for enrollment if it has been planted with an agricultural commodity (including alfalfa and other multi-year grasses and legumes in rotation) in 4 of the 6 crop years prior to 2002 or is located in a designated conservation priority area.[5] In general signups, which account for the vast majority of enrolled acreage, applications from across the country are pooled, ranked, and selected according to an Environmental Benefits Index (EBI) score (see box, "Major Federal Conservation Programs"). Rental payments received by the producer are determined according to county-level average land rents and soil productivity, but producers can bid payments down to increase their EBI score, increasing their chance of selection.

Major Federal Conservation Programs

Conservation Reserve Program (CRP)

Established by the 1985 Food Security Act, the CRP seeks to provide environmental benefits by encouraging the planting of long-term, resource-conserving covers, such as trees or native grasses, on marginal and environmentally sensitive retired agricultural land. In exchange, producers receive an annual rental payment over the term of a 10- or 15-year contract, as well as cost-share payments to offset part of the cost of installing conservation measures.

Producers can enroll land under either a competitive general signup or noncompetitive continuous signup. Under the competitive signup process, each parcel of land proposed for enrollment is assigned an Environmental Benefits Index (EBI) score. The EBI accounts for multiple environmental benefits that would be provided and the expected cost of the contract (USDA-FSA, 2008). Individual scores are calculated for each of the six factors in the EBI: benefits from soil erosion reduction, water quality benefits, wildlife benefits, air quality benefits, cost, and a measure of benefits likely to endure past the contract termination date. Proposals are nationally ranked and selected based on their overall EBI score. Land suitable for certain high-priority conservation practices (e.g., grassed waterways, filter strips, and riparian buffers) and meeting eligibility requirements may be enrolled without being subject to competitive bidding through a continuous signup.

CRP is the largest Federal conservation program; outlays totaled $1.86 billion in 2007 (USDA-FSA, 2008), and projected expenditures for 2008-12 total $10.9 billion.[1]

Environmental Quality Incentives Program (EQIP)

EQIP, established under the Federal Agricultural Improvement and Reform Act of 1996, addresses environmental concerns by providing incentives and technical help to install conservation structures and management practices on working agricultural lands. EQIP pays a portion of the costs for structural and vegetative practices, and most contracts are eligible for cost- share payments up to 75 percent of the costs of conservation structures.

Targeted farmers are eligible for cost-share rates 25 percent higher than the otherwise applicable rate, up to 90 percent. EQIP also provides incentive payments covering 100 percent of the estimated cost of adopting management practices for all farmer types.

Funding decisions are made by States or local entities; each State determines its own conservation priorities and ranking criteria. In some cases, these criteria include targeted farmer status. Prior to 2002, producers could bid down payment rates to increase the probability of funding. The bidding process was eliminated in the 2002 Farm Security and Rural Investment Act, so smaller farmers, who were less able to bid down payment rates, would not be at a competitive disadvantage (USDA, 2003).

The duration of EQIP contracts ranges from 1 to 10 years. Management practice agreements include incentive payments for up to 3 years. A producer can receive up to $300,000 in payments for all contracts held over a 6-year period. Total EQIP funding was $1 billion in 2007, about 22 percent of Federal conservation program funding (USDA-NRCS, 2007); projected expenditures for 2008-12 are $7.2 billion.

Wetlands Reserve Program (WRP)

WRP was established by the 1990 Food, Agriculture, Conservation, and Trade Act. It seeks to restore, protect, and enhance wetlands by retiring marginal agricultural lands. Land enrolled in WRP may be placed in a permanent conservation easement or a 30-year easement or may be improved using a cost-share restoration agreement. Easement payments may not exceed the lowest of the fair market value of the land, a geographic payment cap determined by the Secretary, or an offer made by the landowner.

Applications to WRP are scored based on cost and environmental considerations. The NRCS State Conservationists have the authority to develop scoring criteria and select applications.

WRP is the smallest of the major conservation programs, both in terms of acres enrolled and budget. WRP funding was about $227 million in 2007, or about 6 percent of total Federal conservation program funding (USDA-NRCS, 2007); projected expenditures for 2008-12 are $2.1 billion.

[1] Projected expenditures come from ERS calculations based on the Congressional Budget Office's budget score of the 2008 Food, Conservation, and Energy Act (H.R. 2419).

In WRP, land is eligible for enrollment if it is restorable as a wetland and suitable for wildlife benefits. WRP funds are allocated to State NRCS offices, which set enrollment priorities and make enrollment decisions. In CRP and WRP, enrollment priority is not given to farmers who are new to farming, have limited resources, or are socially disadvantaged. For these programs, participation patterns for all farmers can vary, depending on the relationship between national (in CRP) or State/sub-State (in WRP) conservation priorities and the resource concerns farmers are willing to address.

Like WRP, EQIP funding is allocated to State NRCS offices, and State offices are given the latitude to determine conservation priorities and whether State or sub-State offices make enrollment decisions. Unlike WRP or CRP, some State offices have set aside funds or prioritized enrollments of certain farmer types. The 2008 Farm Act mandates that at least 5 percent of all financial assistance in EQIP be made available to beginning farmers and another 5 percent be made available to socially disadvantaged farmers. The legislation specifies a national threshold and does not require each State or sub-State office to adopt the same threshold.

Targeted farmers qualify for higher payment rates in EQIP. EQIP also includes favorable payment terms to encourage participation by targeted farmers, while CRP and WRP contain no such provisions. In EQIP, all participants can receive financial assistance of up to 75 percent of the cost of installing structural and vegetative practices and 100 percent of the cost of adopting conservation management practices. Since 2002, beginning and limited-resource farmers have been eligible to receive up to 90 percent of the cost of structural and vegetative practices (although not all States offered the higher rates to these farmers). Under the 2008 Farm Act, socially disadvantaged farmers and ranchers became eligible for the higher EQIP rates for these practices, and all three targeted groups became eligible for cost-share rates that were at least 25 percent higher than would otherwise be applicable, up to 90 percent of the cost.

TO WHAT EXTENT DO TARGETED FARMERS PARTICIPATE IN CONSERVATION PROGRAMS?

Many factors can affect farmers' decisions to participate in conservation programs, including whether the financial benefits from enrolling exceed the costs, the type and severity of environmental problems they face, and whether they believe their chances of being accepted will offset the transaction costs of enrolling. Analyzing enrollment patterns can illuminate whether different types of farmers participate in these programs to different degrees.

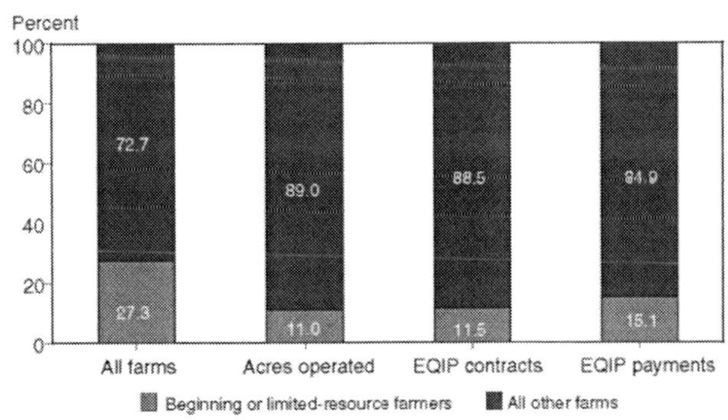

EQIP= Environmental Quality Incentives Program.

Sources: ERS calculations based on USDA's 2007 Agricultural Resource Management Survey Phase III, conducted by the National Agricultural Statistics Service and the Economic Research Service (all farms and acres operated); USDA-NRCS contract data, fiscal 2006 (EQIP contracts and EQIP payments). NRCS contract data identifies beginning and limited- resource farmers, but not socially disadvantaged status.

Figure 4. Share of farms, acres operated, and participation in EQIP, by producer type.

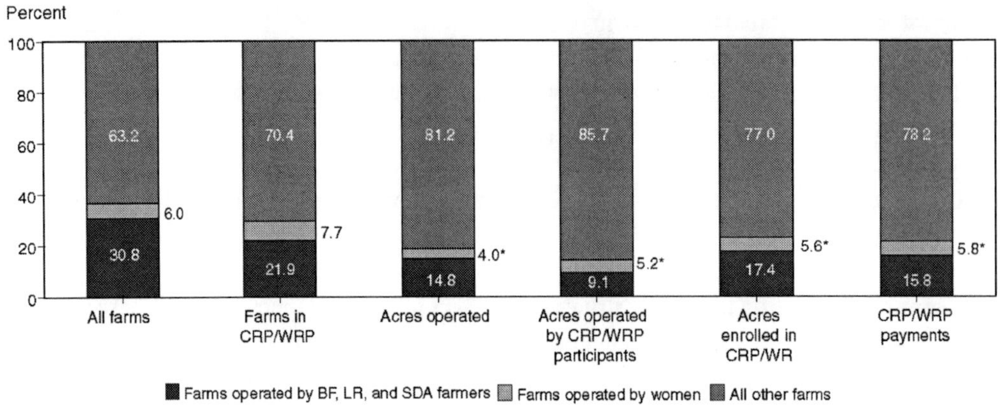

* Coefficient of variation is between 25 percent and 50 percent.

BF=Beginning farmers; LR=Limited-resource farmers; SDA=Socially disadvantaged farmers.

CRP=Conservation Reserve Program; WRP=Wetlands Reserve Program.

Notes: Farms operated by women include farms with female primary operators who do not meet the definitions of beginning, limited-resource, or socially disadvantaged farmers. The "all other farms" category includes farms where women are not primary operators. Totals may not sum to 100 due to rounding.

Source: ERS calculations based on USDA's 2007 Agricultural Resource Management Survey, Phase III, conducted by the National Agricultural Statistics Service and the Economic Research Service.

Figure 5. Share of farms, acres operated, and CRP or WRP participation, by producer type.

Beginning and limited-resource farmers may be under-represented in EQIP, but representation in CRP and WRP is less clear. Participation in EQIP can be measured for farms operated by beginning farmers and limited- resource farmers—the two groups for which data on farmer status were collected during our study period (figure 4).[6] These two groups operated about 27 percent of all farms, but were associated with only 12 percent of EQIP contracts and received 15 percent of EQIP financial assistance payments.[7] Data on the number of acres operated by type of EQIP participant were not available, so evaluating participation by this measure was not possible. Also, although limited-resource farmers were about twice as likely to report farming as their major occupation than beginning farmers, they were no more likely to participate in EQIP.

The extent to which particular farmer types participate in land retirement programs depends on how participation is measured. When participation is measured based on enrolled acres, the data suggest targeted farmers are proportionally represented in these conservation programs. These farmers operated nearly 15 percent of farmland acres and controlled 17 percent of acres enrolled in CRP and WRP in 2007 (figure 5). On the other hand, these farmers operated 31 percent of all farms, but they operated only 22 percent of farms enrolled in CRP or WRP. Taken together, the data suggest that, although disproportionately fewer of these farmers enroll land in land retirement programs, those that do enroll tend to enroll more of their land.

Comparing payment patterns across programs suggests that participation rates by beginning and limited-resource farmers were not clearly higher in EQIP than in CRP/WRP (data on socially disadvantaged farmer participation in EQIP are not available). These two groups of farmers received similar proportions of payments in both programs (15.1 percent in EQIP and 15.0 percent in CRP/WRP). The proportions are similar despite the eligibility of beginning and limited-resource farmers for more favorable payment terms in EQIP since 2002 and their enrollment priority in some States. However, a gap between EQIP and CRP/WRP, in terms of these farmers' participation rates (as measured by payments), could have existed without the favorable terms. A gap might also have existed if a CRP eligibility requirement that all land offered for retirement be previously cropped 4 of the 6 crop years prior to 2002 had not been in effect (USDA-FSA, 2008). A significantly smaller proportion of beginning and limited-resource farmers specialize in crop production (table 2), so fewer of these farmers qualify for CRP enrollment. The relative sizes of the programs may also matter. CRP and WRP expenditures exceeded EQIP expenditures by at least 2 to 1 in recent years, so land retirement programs were able to enroll more participants of all types. A preference for land retirement programs may be due, in part, to beginning and limited-resource farmers finding reduced labor requirements of land retirement practices more appealing (Claassen et al., 2007; Lambert et al., 2006).[8]

Another program design difference between working lands and land retirement programs could affect relative participation patterns across programs. Applicants in CRP and WRP can bid down the amount of payment they are willing to accept to improve their chances of enrollment, but bidding has not been allowed in EQIP since 2002 due to perceptions that it favored larger or more well off farmers (USDA, 2003). Beginning and limited-resource farmers' participation at similar rates in land retirement programs—programs where bidding is allowed—raises questions about whether bidding puts these farmers at a competitive disadvantage. Allowing producers to bid down the payments can improve program cost effectiveness, when enrollment priority is given to producers who could provide environmental benefits most cost effectively. However, many factors affect decisions about participation in working lands and land retirement programs (Lambert et al., 2006), and further studies would need to account for these characteristics to determine whether bidding options do affect participation rates.

Targeted farmers across the United States enroll in conservation programs, but variation across regions suggests national policies could have geographically uneven impacts. In EQIP, beginning farmers held nearly 3,300 (8 percent) of the approximately 39,000 total contracts initiated in 2006. The Southern Plains and Corn Belt regions have the greatest number of EQIP contracts (31 percent of all contracts), but the greatest concentration of beginning-farmer contracts (31 percent of beginning-farmer contracts) is found in the Delta and Mountain regions. Beginning farmers in the Lake States and Southern Plains regions were least likely to participate in EQIP, where they represented 20 percent and 27 percent of the farm population respectively, but held only 3-5 percent of EQIP contracts (figure 6). In Alaska, Hawaii, and the outlying areas in the Pacific Basin and the Caribbean (U.S. territories), at least 35 percent of the contracts were held by beginning farmers. While comprising a large proportion of contracts in these areas, relatively few contracts (186 contracts) were written in these regions, so their impact on beginning farmer participation nationwide is small.

The highest concentration of EQIP contracts with limited-resource producers is located in the Delta region (310 of 1,373 contracts with limited-resource producers initiated in 2006) (figure 7). The U.S. territories also play an important role in limited-resource producer contracts, both in terms of numbers and as a percentage of contracts in the region. In 2006, 174 contracts were written with limited-resource farmers in the Pacific Basin and Caribbean, and these contracts represent over 50 percent of all contracts held by farmers in the U.S. territories.

EQIP PARTICIPATION ON AMERICAN INDIAN RESERVATIONS

Although much of this chapter focuses on the enrollment patterns of different types of farmers, it is also important to examine conservation programs in certain types of communities or geographic areas. This places more emphasis on where conservation practices are adopted and what types of communities may accrue environmental benefits from conservation programs.

EQIP participation and farm characteristics, by location on or off American Indian reservations or tribal trust land

EQIP participation	Contracts on reservation farmland	Contracts off reservation farmland[1]	All contracts[2]
Number of contracts	641	14,707	39,030
Percent of contracts associated with BF or LR	21.2	10.8	11.5
BF or LR contracts:			
Median total contract payment ($)	28,428	17,365	8,618
Percent of total payments to BF or LRP contracts	1.7	13.5	15.1
Contracts with other farmer types:			
Median total contract payment ($)	18,048	12,144	12,245
Percent of total payments	4.5	80.3	84.9
Selected farm characteristics[3]	Farms operating reservation farmland	Farms not operating on reservations[4]	All farms
Average acres operated	479*	592	582
Percent of total acres operated	7.7	92.3	100
Percent of farms	9.3	90.7	100
Average value of production ($)	80,415*	167,803	159,682

EQIP=Environmental Quality incentives Program; BF = Beginning farmer; LR = Limited resource farmer.

*Indicates significant differences between reservation and nonreservation farms based on pairwise two-tailed delete-a-group jackknife t- statistics at the 90-percent confidence level. See Kott, 2001.

[1] Includes contracts in States where there are reservation contracts: Arizona, California, Colorado, Idaho, Kansas, Michigan, Minnesota, Montana, Nebraska, Nevada, New Mexico, North Carolina, North Dakota, Oklahoma, Oregon, South Dakota, Utah, Washington, Wisconsin, and Wyoming.

[2] Includes all contracts in all States from 2006.

[3] Calculated using the 2007 Agricultural Resource Management Survey (ARMS) Phase III.

[4] The comparison group for ARMS is not directly comparable with the EQIP contract data, although ARMS is designed to provide nationally representative estimates. Includes farms in the following States: Arizona, Florida, Idaho, Kansas, Michigan, Minnesota, Montana, Nebraska, Nevada, North Dakota, Oklahoma, Oregon, South Dakota, Utah, Washington, Wisconsin, and Wyoming.

Source: Economic Research Service.

As an example, we examined EQIP participation and practice adoption on American Indian reservations and trust land. Reservation communities may be considered disadvantaged by a variety of measures, including household income and poverty rates, educational attainment, health outcomes (Ward et al., 2004), and food security (Gundersen, 2008). Whether this characterization extends to the environmental benefits derived from EQIP is not clear; the characteristics that define a community as disadvantaged may not be closely associated with factors that determine EQIP participation and practice adoption. For example, farmers on reservations are not necessarily disadvantaged; reservation farmland may be leased to non-American Indian farm operators, and farm operators who are American Indian may not share the characteristics of other nonfarmers in the community.

By comparing EQIP administrative data with the established borders of reservations from the Bureau of Indian Affairs, we determined which EQIP contracts were within the borders of American Indian reservations.

Nationwide, 641 contracts in 2006 were on a reservation, or about 2 percent of all contracts. We then compared these reservation contracts with contracts outside American Indian reservation borders.

Contracts on reservations received larger total payments from EQIP contracts than contracts off reservations. The median total contract payment was between $6,000 and $11,000 larger on reservations, even after accounting for differences between beginning and limited-resource farmers and other farmers. Beginning and limited- resource farmers are eligible for higher cost-share rates, which may result in larger payments for the same size practice. In this case, the scale of practices may account for a portion of the larger payments on reservations. Calculating the average sizes of the most common structural practices (e.g., fences and pipelines) confirm that contracts on reservations tend to involve larger structures. Some evidence suggests that larger payments and practices may yield greater environmental benefits, at least when addressing certain environmental concerns (Wu et al., 2004). To the extent that this relationship holds for EQIP participants, per contract environmental benefits may be greater for reservation contracts.

Farms on reservations are less likely to be enrolled in EQIP than nonreservation farms. Reservations account for about 9.3 percent of farms and 7.7 percent of operated farmland in the comparison States, but about 4.2 percent of EQIP contracts and 6.2 percent of EQIP funding in 2006. Participation in EQIP depends on a variety of landscape and farm-level factors, so it is difficult to say if under-representation in this case amounts to being underserved. For example, contracts on reservations are less likely to be associated with cropland that is designated as highly erodible. Under-representation in EQIP may reflect differences in environmental conditions that make participation less likely on reservation farms.

Are American Indian reservations less likely to accrue benefits from participation in EQIP? The results suggest a mixed conclusion. Reservation farms appear to be under-represented in EQIP, but this fact alone is not conclusive. EQIP contracts on reservation farmland are more likely to be associated with beginning or limited- resource farmers and to be eligible for higher cost-share rates. The higher cost-share rates may encourage participation among farmers who might not otherwise be able to participate in EQIP. Also, the typical contracts in EQIP are larger and may provide greater environmental benefits per contract on reservations. Contracts on reservation farmland also appear to address different environmental issues, which suggests that the benefits derived from EQIP contracts are different for reservation communities. Reservation contracts are less likely to address soil erosion and water quality problems and more likely to address issues related to plant condition and domestic animals. Addressing different environmental concerns may also affect the cost and the size of practices. A variety of factors, however, infl uence the benefits derived from conservation practices, and further study is needed to examine how these factors may be related to the distribution of environmental benefits from EQIP.

In CRP and WRP, 18 percent of all contracts were held by beginning farmers during the 2005-07 period, and these farmers participated in proportions similar to that of the farm population in a few regions (figure 8). In the Delta region, land retirement programs were popular with beginning farmers. These farmers represented 17 percent of the farmer population and operated about 14 percent of the acres, but accounted for 23 percent of CRP/WRP participants and about 35 percent of acres enrolled in the region.

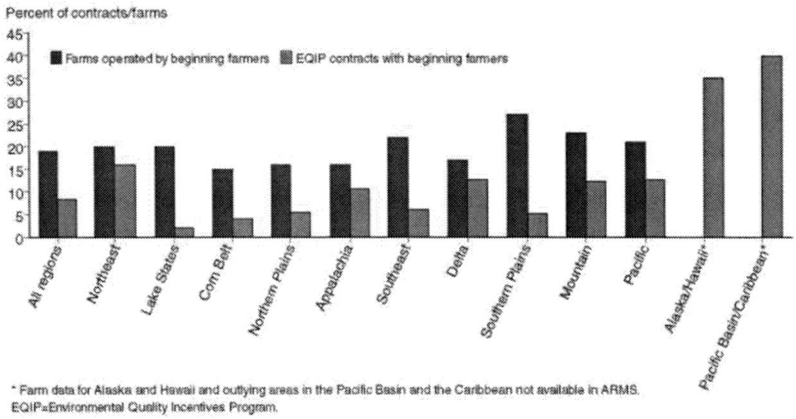

Sources: ERS calculations based on USDA's pooled 2005-07 Agricultural Resource Management Survey, Phase III, conducted by the National Agricultural Statistics Service and the Economic Research Service (farm population); USDA-NRCS contract data, fiscal 2006 (EQIP contracts).

Figure 6. Beginning farmers in the farm population and as a share of EQIP contracts, by region.

In land retirement programs, notable differences also exist in regional participation patterns for limited-resource and socially disadvantaged farmers. During 2005-07, these farmers in the Appalachian and Southern Plains regions were more likely to be enrolled in CRP/WRP (figure 9). About 28 percent of CRP/WRP participants in these regions are limited-resource or socially disadvantaged farmers, but they comprise only 16-22 percent of the farmer population. The Southern Plains region includes States with a fair number of CRP contracts, so in addition to a high percentage of enrollments, limited-resource producers are likely to hold a large number of CRP contracts in this region (USDA-FSA, 2008).

The geographic distribution of targeted farmers may be most relevant when analyzing the impact of nationally specified program requirements. For example, the "conservation access" provision of the 2008 Farm Act that establishes a 5-percent set aside of EQIP funds for beginning and socially disadvantaged farmers is a national threshold, but the provision may have a larger impact on enrollment if administered at the regional level. Nationwide, beginning farmers received more than 10 percent of EQIP funding over the 2004-06 period, but within a few regions, payments to beginning farmers fell below the 5-percent threshold.

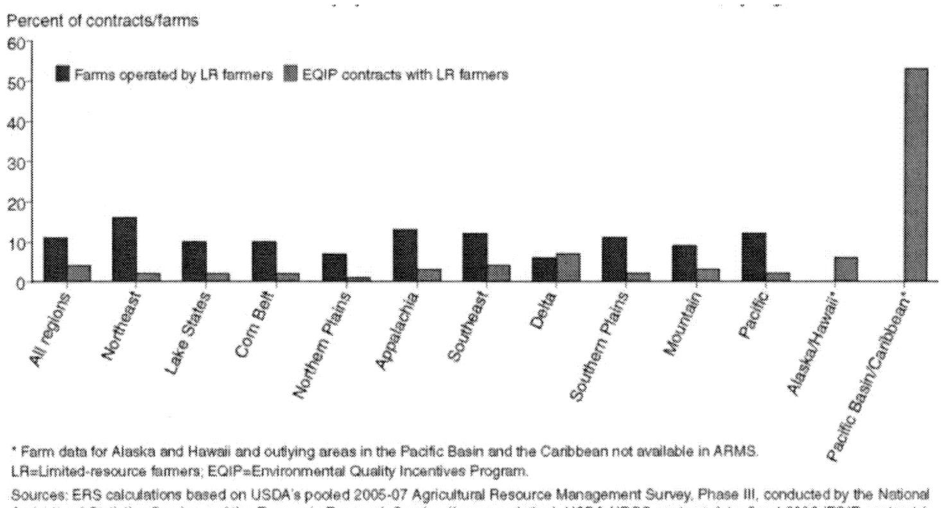

Sources: ERS calculations based on USDA's pooled 2005-07 Agricultural Resource Management Survey, Phase III, conducted by the National Agricultural Statistics Service and the Economic Research Service (farm population); USDA-NRCS contract data, fiscal 2006 (EQIP contracts).

Figure 7. Limited-resource farmers in the farm population and as a share of EQIP contracts, by region.

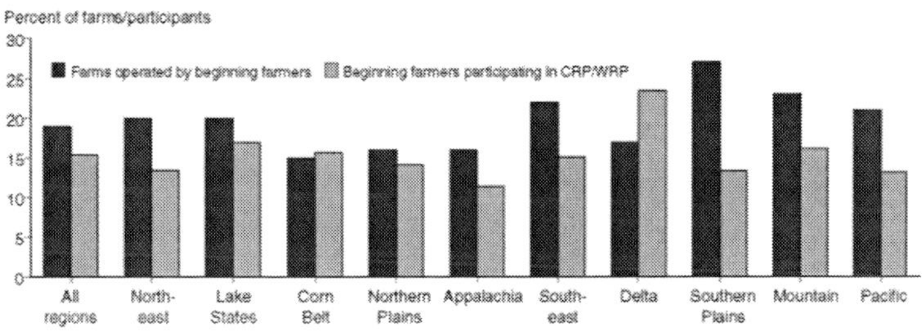

CRP=Conservation Reserve Program; WRP=Wetlands Reserve Program.

Source: ERS calculations based on USDA's pooled 2005-07 Agricultural Resource Management Survey, Phase III, conducted by the National Agricultural Statistics Service and the Economic Research Service.

Figure 8. Beginning farmers/ranchers as a share of farm population and CRP/WRP participants, by region.

How Do Targeted Farmers Use Conservation Programs?

Producers who participate in conservation programs employ a variety of practices to mitigate environmental problems that arise as a byproduct of agricultural production. Producers will choose the problems they want to address and practices they want to implement that best meet their needs, provide the greatest financial returns, and are likely to receive funding from conservation program managers. Examining natural resource characteristics and enrollment data highlight whether targeted farmers participate in conservation programs differently than other farmer types.

EQIP

Do beginning and limited-resource farmers participating in EQIP operate more environmentally sensitive land? Some evidence suggests "yes." While precise data on the underlying physical characteristics of land receiving treatment through EQIP-funded practices are not available, we can characterize two aspects of the land in the general vicinity of EQIP contracts: soil erodibility and proximity to quality-impaired waters.[9] When examining correlations between producer types and highly erodible soils, we limited our analysis to four States—Louisiana, Arkansas, Mississippi (States in the Delta region), and Alabama—because of the computational burden involved in associating soils data with EQIP contracts. These States were selected because they have both a relatively large number of EQIP contracts (each had more than 1,000 contracts signed in 2006) and a minimum of about 5 percent of contracts each with beginning and limited-resource producers.

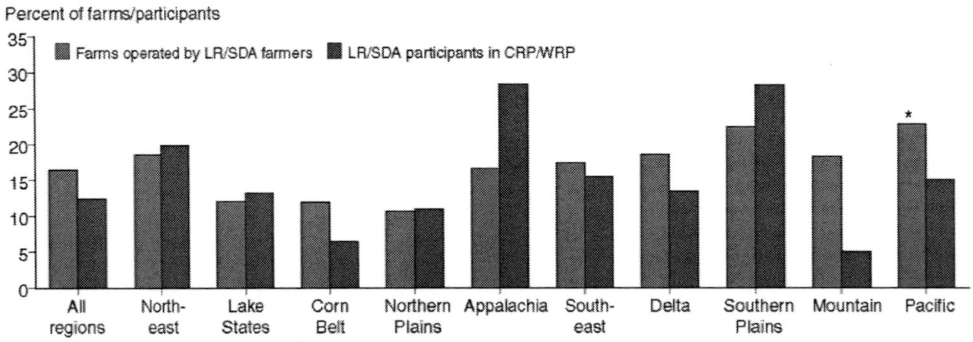

* Coefficient of variation is between 25 percent and 50 percent.

LR=Limited-resource farmers; SDA=Socially disadvantaged farmers.

Source: ERS calculations based on USDA's pooled 2005-07 Agricultural Resource Management Survey, Phase III, conducted by the National Agricultural Statistics Service and the Economic Research Service.

Figure 9. Limited-resource and socially disadvantaged farmers as a share of farm population and CRP/WRP participants, by region.

When examining the relationship between EQIP contracts and quality- impaired waters, we used national-level data. We determined proximity of farms to streams and water bodies with pollution problems by using hydrologic data from the U.S. Environmental Protection Agency's database on waters considered impaired under the Clean Water Act. These distance measures were used for all EQIP contracts in the continental United States.

Using 2006 EQIP contract data, our analysis reveals:[10]

- In the Delta region, beginning-farmer contracts were located in areas with significantly greater average percentages of highly erodible land (54 percent) compared with contracts with other farmer types (47 percent).

- For limited-resource producers in the Delta region, an average of 50 percent of nearby land was highly erodible, which was not statistically different from the percentage for other farmer types.
- Nationally, both beginning and limited-resource farmer contracts tended to be located in areas significantly closer, on average, to impaired waters than contracts with other farmer types.

These analyses suggest that of those with EQIP contracts, beginning farmers in the Delta region may be located in areas that are more marginally productive compared with other farmer types, but that limited-resource farmers may not necessarily operate in areas with more erodible soils. Limited-resource farmers may be more likely to occupy more productive land than beginning farmers because a larger percentage of limited-resource farmers report farming as their major occupation and thus may seek to farm less highly erodible land. Producer characteristics and the underlying qualities of the land they operate can vary substantially across regions of the country, so additional analysis could determine if the correlations with erodible land in the Delta region are representative of the United States. The national analysis of a farm's proximity to impaired waters revealed that both beginning and limited-resource farmers may be operating in more environmentally sensitive areas than other farmers.

Beginning and limited-resource farmers participating in EQIP have different conservation priorities from other participants. When different types of producers face different environmental problems, the greatest conservation program benefits can be reaped when their particular problems are treated. The resource concerns treated with EQIP funds depend on the producer's willingness to treat the problem, as well as government priorities. Eight broad categories of resource issues are addressed through EQIP:

1. Air quality.
2. Domestic animal needs.
3. Fish and wildlife.
4. Plant productivity/quality.
5. Soil condition.
6. Soil erosion.
7. Water quality.
8. Water quantity.

Of these, the four most frequently addressed issues are domestic animal needs, plant productivity/quality, soil erosion, and water quality, which are treated by more than 85 percent of all EQIP contracts.

Of the four most commonly addressed resource issues, beginning and limited- resource farmers enrolled in 2006 were more likely to address plant productivity/quality issues and domestic animal forage and health needs compared with other participating farmer types (table 3). Limited-resource farmers appeared less likely to address water quality problems compared with beginning farmers or other farmer types, even though our analysis revealed that they tended to be located closer to quality-impaired waters than other program participants. Only 25 percent of limited-resource farmers address water quality, while 36-3 8 percent of beginning and other farmer types do.

Table 3. Resource concerns addressed with EQIP contracts, by farmer status, 2006.

Concern[1]	Beginning farmer/rancher	Limited-resource farmer/rancher	Other farmer types[2]
Percent of EQIP contracts addressing concern			
Domestic animal forage and health needs	36	32	22
Plant productivity/quality	44	38	34
Soil erosion	34	38	37
Water quality	38	25	36
Resource concerns per contract[3]	1.98	1.98	1.7

EQIP=Environmental Quality Incentives Program.

[1] Remaining four resource concerns not listed are air quality, fish and wildlife, soil condition, and water quantity. None of these four issues are treated by more than 10 percent of contracts for each group.

[2] Includes all contracts held by farmers who are neither beginning nor limited-resource farmers or ranchers.

[3] Counts are not weighted by the number of practices implemented to treat the resource concern. If treated by multiple practices, the resource concern is counted once to eliminate the effect of multiple practice adoption choices.

Source: ERS analysis of USDA-Natural Resources Conservation Service contract data, fiscal 2006.

These data suggest that farmer groups have different priorities for treating resource problems. They also suggest that a closer look at whether particular constraints prevent limited-resource farmers from treating water quality problems may be warranted. A similar proportion of beginning-farmer contracts treated water quality problems compared with other farmer types in EQIP, even though other farmer types are larger and face more regulatory requirements for reducing water pollution under the Clean Water Act.

Beginning farmers had more expensive EQIP contracts compared with other participating farmer types, while limited-resource farmers had less expensive contracts. Beginning and limited-resource farmers in EQIP have somewhat different resource priorities than other participating farmer types, and costs differ as well. On average, beginning farmers enrolled in EQIP had significantly more costly contracts overall ($33,682 compared with $28,883, including both producer and government share of the costs) and received larger EQIP payments (the government share of the costs was $26,646 compared with $18,056) (table 4). Because EQIP funds a larger portion of the costs for beginning farmers to install structural practices, such as fences, waste-handling facilities, and vegetative buffers, larger payments to these producers are not surprising. Beginning farmers, however, are more likely to operate residential, retirement, or smaller farms, and a smaller proportion specialize in livestock production compared with other farmer types (see tables 1 and 2)—all characteristics that could contribute to smaller payment sizes. Livestock producers often install more expensive conservation practices to treat animal-related resource problems.

Unlike beginning farmers, limited-resource farmers in EQIP tend to have less costly contracts overall compared with other farmer types. Limited-resource farmers also tend to have smaller farms, on average, in addition to being the most likely group of farmers to face liquidity constraints (see table 2).

Table 4. Average EQIP contract and practice costs, by farmer status, 2006.

Item	Beginning farmer/rancher		Limited-resource farmer/rancher		Other farmer types[1]
	Value	Difference from other farmer types	Value	Difference from other farmer types	
All practices:					
Total cost of contracted practices (average $)[2]	33,682	4,799**	23,242	-5,641**	28,883
Total government payments (average $)	26,646	8,590**	19,175	1,119	18,056
Total producer expenditures (average $)	7,035	-3,791**	4,067	-6,759**	10,827
Practices per contract	4.44	1.4**	3.70	0.66**	3.04
Cost-share practices:					
Total cost per practice (average $)[2]	5,865	-1,514**	5,008	-2,370**	7,378
Government payment per practice (average $)	4,545	390**	4,053	-101	4,155
Producer expenditures per practice (average $)	1,320	-1,904**	955	-2,269**	3,224
Practices per contract	3.64	0.95**	3.10	0.41**	2.69
Share rate (average)	0.76	0.19**	0.82	0.25**	0.57
Management practices:					
Government payment per practice (average $)	1,022	-856**	971	-907**	1,878
Practices per contract	1.79	0.19**	1.63	0.03	1.60

EQIP=Environmental Quality Incentives Program.

** Column differences that are significant at the 95-percent level using a two-tailed t-test. Practices per contract are calculated for each group by dividing the total number of unique practices by the total number of contracts.

[1] Includes all contracts held by farmers who are neither beginning nor limited-resource farmers or ranchers.

[2] Includes government payments and producer expenditures for each practice.

Source: ERS analysis of USDA-Natural Resources Conservation Service contract data, fiscal 2006.

Beginning and limited-resource farmers received smaller per practice EQIP payments and implemented smaller scale practices but adopted a greater number of practices. Beginning and limited-resource farmers differ from other farmer types in terms of the sizes and types of practices funded by EQIP. The average incentive payment per management practice was significantly smaller for beginning and limited-resource farmers (table 4), due to implementation of management practices on a smaller scale. For example, beginning farmers implemented nutrient management (the most common management practice) on about 60 percent fewer acres, on average, compared with other participating farmer types. Beginning farmers also appear to implement this practice at a greater cost per acre.

In contrast to incentive payments for management practices, government payments for structural (cost-share) practices were larger for beginning farmers than for other participating farmer types, even though beginning farmers tended to implement these practices on a smaller scale. Beginning farmers are paid at higher cost-share rates, so the Federal Government picks up a higher portion of the total practice cost (see table 3). An analysis of structural practice sizes by farmer type confirms that, for two of the most common structural practices (e.g., fences and pipelines), beginning farmers installed significantly fewer feet of the structures. While higher per practice payments suggest that beginning farmers may not be treating environmental problems cost effectively, they may also signal that beginning farmers face more serious problems that need treatment. Both beginning and limited- resource farmers implemented significantly more cost-shared practices within a single contract—3.6 and 3.1 cost-shared practices on average, compared with 2.7 cost-shared practices by other farmer types.

A policy designed to pay a larger portion of the costs of structural and vegetative practices for typical targeted farmers could ease financial constraints to adopting certain practices. Such a policy, however, may encourage these farmers to adopt more structural and vegetative practices than they would otherwise, simply because they face a lower out-of-pocket cost per unit of practice installed. While the policy could be a contributing factor, a variety of other factors may explain the higher incidence of cost-shared practices in beginning and limited-resource farmer contracts. For example, differences in production and household characteristics could matter. Previous research has found that the number of structural practices installed was positively associated with land ownership (Lambert et al., 2006), and a larger percentage of beginning and limited-resource farmers tend to be full owners of the land they operate (see table 2). Differences in opportunities to receive funding for practice installation through other national and local government programs may also affect practice adoption patterns in EQIP (Feng et al., 2006; Feng et al., 2005).

Historically, not all States offered higher cost-share rates to beginning and limited-resource farmers. For example, Oregon, Missouri, and Iowa did not pay higher cost-share rates to beginning or limited-resource producers in 2006. However, the 2008 Farm Act provision mandating that all targeted farmers receive a cost-share rate at least 25 percent higher than the normal rate in EQIP (up to a maximum share rate of 90 percent) may increase their participation, particularly in States that previously offered them no cost-share advantage.

In addition to a greater number of practices, beginning and limited-resource farmers have increasingly adopted cost-shared structural practices in combination with management practices. The proportion of beginning farmers using combinations of structural and management practices in EQIP increased from 43 percent to about 51 percent during 2004-06 (figure 10). The proportion of beginning farmers implementing strictly cost-share practices through EQIP declined from 52 percent to 42 percent over the same period. Practice adoption patterns were similar for limited-resource farmers. The practice combinations adopted by other participating farmer types, however, remained steadier over this period, with about 31 percent adopting a combination of structural and management practices and 50 percent installing strictly structural practices.

Beginning farmers are more likely to have longer contracts than other program participants. Other farmer types participating in EQIP are more likely to have single-year contracts (40 percent compared with 27 percent for beginning farmers). Over the 2004-06 period, more beginning farmers signed EQIP contracts that lasted at least 4 years (up to 27 percent of farmers in 2006), whereas only 17 percent of other farmer types had contracts that long. Although the 2002 Farm Act legislation reduced the minimum length of EQIP contracts from 5 years to 1 year, the reduction in length appears to be less important for beginning farmers. Shorter contract length can reduce risks from having long-term obligations to fulfill, as well as uncertainty about farm returns, which could be considered attractive benefits to all farmer types (USDA, 2003). Spreading implementation of certain practices over more contract years, however, may suit farmers who face labor or capital constraints that pose difficulties for rapid adoption.

Taken together, these findings reveal that beginning and limited-resource farmers participate in working-lands programs in different ways than other farmer types. Beginning and limited-resource farmers participating in EQIP may be located in areas that are more environmentally sensitive and, in the Delta region, more marginally productive compared with other program participants, and they address somewhat different resource issues at different costs.

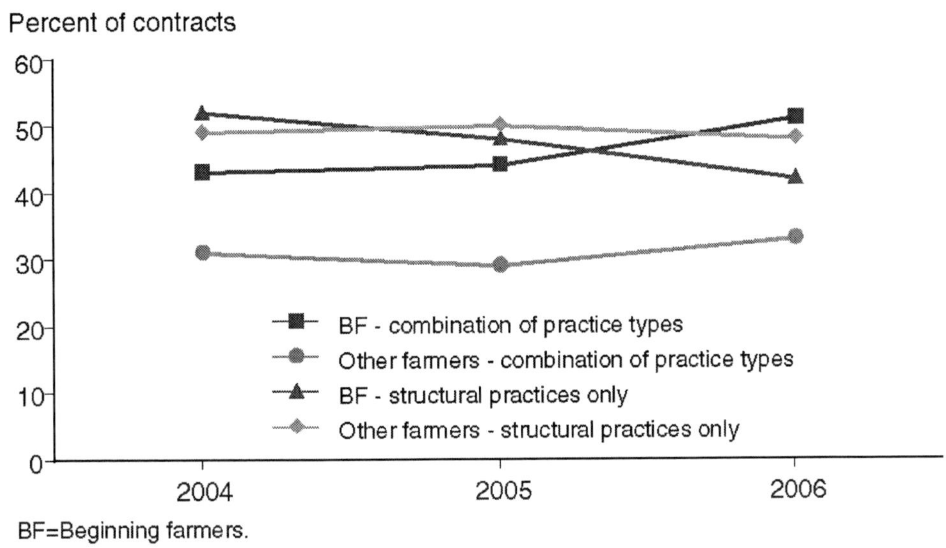

BF=Beginning farmers.
Source: ERS analysis of USDA-NRCS contract data, fiscal 2006.

Figure 10. Types of practices implemented in EQIP, by beginning farmer status, 2006.

CRP and WRP

Analysis of environmental problems, resources receiving treatment, and costs incurred by participants suggest that farmer participation in land retirement may not follow the same pattern as in EQIP. CRP and WRP involve a different set of program incentives that may affect the pattern of enrollment choices.

Targeted farmers participating in CRP/WRP tend to be concentrated in counties with the highest average percentage of cropland designated as highly erodible. Other program participants are more likely to be located in counties with a lower percentage of land that is highly erodible. This distribution of farmers is similar to that between beginning farmers and other program participants in EQIP in the Delta region. Contract data do not identify producer status in land retirement programs, so only general characterizations are possible using ARMS and the 1997 National Resources Inventory data on the erodibility of land.

Targeted farmers who participate in CRP may be more likely to treat soil erosion and water quality problems that provide more environmental benefits. Targeted farmers participating in CRP tend to be concentrated in counties where participants, in general, were expected to generate the highest soil erosion reduction and water quality benefits. Approximately 44 percent of targeted farmers who participate in CRP/WRP are located in counties where participants as a whole generate the highest third of Environmental Benefit Index (EBI) scores for average soil erosion reduction benefits.[11] About 33 percent of other participating farmer types are located in those counties. Similarly, about 40 percent of targeted participants in CRP/ WRP farm in counties with the highest third of EBI scores for average water quality benefits, compared with about 29 percent for other farmer types.[12]

Contract costs and practices adopted by targeted farmers participating in CRP and WRP share some similarities to contracts with other farmer types, but differences exist among particular groups. Beginning farmers received land retirement payments that were about $1,400 less than payments to other farmer types. This payment difference is statistically significant, but the number of acres enrolled was not significantly different.[13] The opposite holds for limited-resource and socially disadvantaged farmers; average payments were not significantly different from those of other farmer types, but the average total acres enrolled is significantly larger (table 5). Differences in payment and enrolled acreage patterns could exist due to differences in characteristics between these groups of farmers (tables 1 and 2). Unlike EQIP, differences in CRP/WRP payments across producer types are not a result of special treatment, such as the differential cost-share rates offered to targeted farmers in EQIP.

Table 5. CRP/WRP payments and acres enrolled, by farmer status, 2005-07[1].

Item	Beginning farmer/rancher	Limited-resource or socially disadvantaged farmer/rancher	Other farmer types[2]
CRP/WRP payments:			
Total payment received (average $)	4,645*	4,734	5,851
Payment per acre enrolled (average $)	61.1 *	61.2*	75.6
CRP/WRP acreage:			
Total acres enrolled (average)	116.5	102.5	125.8
Share of acres operated enrolled in CRP/WRP (average percent)	54.2*	57.1*	35.5

CRP=Conservation Reserve Program. WRP=Wetlands Reserve Program.

* Significant differences compared with "other farmer types" at the 95-percent level.

[1] Observations from 2005-07 are pooled to increase sample size. Includes acres and payments from the Conservation Reserve Enhancement Program (CREP).

[2] Includes those who are not targeted farmers or ranchers. Qualitative results do not change when women are included in the definition of socially disadvantaged.

Source: ERS calculations based on USDA's pooled 2005-07 Agricultural Resource Management Survey, Phase III version 1, conducted by the National Agricultural Statistics Service and the Economic Research Service.

On a per acre basis, targeted farmers received significantly lower CRP/WRP payments, relative to other participating farmer types (table 5); per acre payments were between 25 percent and 30 percent lower for these farmers. Several characteristics could contribute to this per acre difference, including concentrations of these farmers in the Southern Plains where payment rates are typically low (Ahearn and Newton, 2009). The differences could also arise due to underlying infl uences on bidding behavior. CRP and WRP allow landowners to bid down the rental payment they are willing to accept to improve their chances of enrollment. Many targeted farmers do not farm as a primary occupation and rely more heavily on off-farm sources of income (table 2). These farmers, in particular, may be willing to bid down the rental payment more than other farmer types.

Table 6. Practice adoption in CRP and WRP, by farmer status, 2001.

	Targeted farmers (A)	Other farmer types (B)	Difference (A - B)
Frequency of practice adoption:		*Percent of farms*	
Whole-field enrollment	70.5	63.6	6.9
Grass buffers or waterways	43.9	26.8	17.1
Wildlife habitat	41.3	14.5	26.9
Other (riparian tree buffers, wetland restoration)	5.9	12.6	-6.7
Number of practices adopted:			
0	17.6	19.3	-1.7
1	35.1	53.5	-18.4
2	15.4	20.2	-4.8
3	31.9	4.4	27.5
4	0.0	2.6	2.6

Notes: No differences are significant at the 95-percent level using the delete-a-group variance calculation (Kott, 2001). Totals may not sum due to rounding.

Source: ERS calculations based on USDA's 2007 Agricultural Resource Management Survey, Phase III, conducted by the National Agricultural Statistics Service and the Economic Research Service.

Targeted farmers enrolled a larger portion of their operated acreage in CRP or WRP. The average beginning farmer had more than 54 percent of operated acreage enrolled in CRP or WRP; the average limited-resource or socially disadvantaged farmer enrolled 57 percent (table 5). A greater percentage of targeted farmers tended to be full owners and may find it easier to make decisions to retire larger amounts of land under CRP/WRP contracts of 10 years or more (Lambert et al., 2007). Indeed, previous research found tenure is positively associated with the proportion of farm acreage enrolled in land retirement programs (Lambert et al., 2006). Also, many of these farmers are retired and may be more likely to enroll land in land retirement programs (Wu, 2004).

Targeted farmers do not implement significantly different types or numbers of practices in land retirement programs. Whole-field practices were the most common practices among all participants in 2001, according to the most recent data available that allows analysis of practice adoption by producer type. A larger percentage of targeted farmers retired whole fields and rehabilitated land for wildlife needs than other farmer types, and a larger percentage adopted a greater number of the listed practices, but these differences were not statistically significant (table 6).[14]

Conclusions and Policy Implications

Since 2002, some Federal conservation programs have included provisions aimed at encouraging participation by targeted farmers. Little is known, however, about whether changing the participation rates of targeted farmers might affect benefits generated by, or the cost effectiveness of, conservation programs. To improve understanding about the impact on program outcomes from improving accessibility for targeted farmers, we analyzed participation patterns in EQIP and CRP/WRP to identify whether targeted farmers differ from other farmers in the types of conservation benefits they provide and the costs of doing so. Our analyses suggest that targeted farmers may play an important role in conservation, and they participate in conservation programs in different ways from other farmers.

By several measures, beginning and limited-resource farmers appear less likely to participate in EQIP relative to their prevalence in the farmer population as a whole. Targeting these farmers for participation could provide them with greater access to conservation program funding. In land retirement programs, the relationship is less clear. Relative to their proportion in the general farm population, a smaller proportion of farms operated by targeted farmers participated in CRP and WRP. The proportion of enrolled land controlled by these farmers, however, was similar to the proportion of all farmland they operate.

Understanding participation patterns across working lands and land retirement programs is also important, particularly since the 2008 Farm Act increased funding for working-lands programs, but reduced the amount of land that could be enrolled in the CRP. If socially disadvantaged farmers are similar to beginning and limited-resource farmers and are less likely to enroll in EQIP, then a shift in conservation support to working-lands programs could reduce the overall access targeted farmers have to conservation funding. Data on socially disadvantaged farmer participation in EQIP would be needed to more definitively address this issue.

Some evidence suggests that targeted farmers participating in conservation programs, as a group, may face more pressing environmental problems than other farmers. In the Delta region, beginning farmers may be able to provide more conservation benefits than other farmer types because they are more likely to farm more erodible land. Encouraging participation by such farmers in regions where this relationship holds could increase the environmental benefits provided by conservation programs.

Choices about which environmental problems to address and which practices to adopt ultimately determine the benefits and costs of program participation. For example, limited-resource farmers are less likely to adopt practices in EQIP that address water quality problems, although they are more likely to be located in watersheds with impaired water quality. In CRP and WRP however, targeted farmers are concentrated in counties with more highly erodible cropland, and they are also concentrated in counties that provide the greatest expected levels of soil erosion reduction and water quality improvement benefits.

While targeted farmers may provide more of some benefits, it is not clear that these farmers can provide benefits more cost effectively than other farmers. Beginning farmers in

EQIP tended to receive larger payments per practice, to implement smaller practices, and to incur larger per acre practice costs for nutrient management (the most common management practice) relative to other farmers. Yet, beginning and limited-resource farmers adopt greater numbers of practices in EQIP, and they have increasingly adopted cost-shared structural practices in combination with management practices. Research suggests that combinations of practices are more effective at providing environmental benefits than single practices, implying that, even if their treatment costs per acre are higher, beginning and limited-resource farmers are not necessarily providing fewer benefits than other farmers (Lerch et al., 2005; Berry et al., 2003).

Program costs per enrolled acre in CRP and WRP were smaller for targeted farmers. If lower per acre payments in CRP/WRP imply that targeted farmers may provide environmental benefits more cost effectively, targeting enrollment in land retirement programs to these producers could increase program cost effectiveness. Our analysis of costs and the distribution of contracts relative to soil erodibility suggest a closer look at the issue may be warranted. Land retirement contract data that included identifiers of farmer type along with benefits data could assess whether farmers with low-cost contracts generate the same or greater levels of benefits.

Designing programs to be broadly accessible can increase the flexibility program managers have to achieve program goals cost effectively. Quantifying the environmental benefits generated by participating farmers would be needed to determine if targeting farmers (beginning, limited resource, or socially disadvantaged) provides benefits more cost effectively.[15] This analysis finds that these farmers participate in different ways than other farmers, leaving open the possibility that encouraging targeted farmers to enroll in conservation programs could result in tradeoffs with cost-effective provision of environmental benefits. These types of farmers, however, tend to operate fewer acres, so an important consideration is whether differences between these and other farmer types are due more to smaller farm size than to farmer type.

APPENDIX. DATA SOURCES

This research relies on the observation of farmer and rancher participation in conservation programs and on the ability to distinguish between different types of farmers and ranchers. We used data from several USDA sources, including annual Agricultural Resource Management Survey (ARMS), Environmental Quality Incentives Program (EQIP), and Conservation Reserve Program (CRP) contract databases.

ARMS. USDA's Agricultural Resource Management Survey is the only annual source of data for crop and livestock production practices, farm business financial information, and household and operator characteristics for a nationally representative sample of farmers and ranchers in the continental United States. We primarily used the 2007 ARMS (Phase III) to identify the geographic distribution of farms and ranches, their participation in Federal land retirement programs, and the characteristics of operators and households. For some analyses, ARMS data from 2005 and 2006 are pooled with 2007 data to increase sample size and improve the reliability of estimates.[16] Information on practices applied in CRP and Wetlands

Reserve Program (WRP) uses 2001 ARMS data on conservation practices adopted by corn farms (the most recent year available). For a more detailed description of ARMS data, see http://www.ers.usda.gov/briefi ng/ARMS.

There are a few limitations to using ARMS data:

1. Not all producers surveyed in ARMS receive the same questionnaire, so some information – including separate payments received for CRP and WRP participation – is available for only part of the surveyed population. For some analyses, the limited number of useable observations prevents analyses at regional or subregional scales.
2. Operators with certain household characteristics may be under- sampled in ARMS. Such undersampling may preclude statistically reliable conclusions about targeted farmers and ranchers in some cases (data in tables are notated where large coefficients of variation suggest that reliability is suspect).
3. ARMS collects information on CRP/WRP participation, but it does not distinguish an operator with one CRP/WRP contract from an operator with multiple contracts. As a result, ARMS data will undercount conservation program contracts held by particular farmer types if participants hold several CRP or WRP contracts.

EQIP contract data. Information about each EQIP contract is housed in the Protracts database maintained by USDA's Natural Resources Conservation Service (NRCS). Protracts contains information about the conservation practices that farmers and ranchers agree to apply to the land, government cost-share rates and payment amounts (including incentive payments), the resource concerns that practices address, and beginning and limited-resource farmer/rancher status. We used EQIP data to review contracts for which funds were obligated in fiscal years 2004-06. These data include all EQIP practices contracted for these years, regardless of when they were applied or when payment occured. Participants' status as socially disadvantaged was not recorded in the Protracts system over the 2004-06 period.

CRP contract data. CRP contract data is maintained by USDA's Farm Service Agency. These data do not include indicators of targeted farmer status, but information on location (county) and Environmental Benefits Index (EBI) scores are useful for characterizing the geographic distribution of EBI scores relative to the geographic distribution of targeted farmer participation data from ARMS.

2007 Census of Agriculture data. The Census of Agriculture is conducted by USDA's National Agricultural Statistics Service every 5 years. It is a complete count of U.S. farms and ranches and farm operators. The census contains information on land use and ownership, operator characteristics, production practices, income and expenditures, and more. These data are useful for summaries at a county or other aggregate level. The data include characteristics identifying certain farmer types, including beginning farmers and some groups classified as socially disadvantaged farmers. It does not allow us to identify all socially disadvantaged groups or farmers defined as having limited resources. For more information about Census of Agriculture data, see http://www.agcensus.usda.gov/.

National Resources Inventory data. The National Resources Inventory (NRI) is a survey conducted by USDA's Natural Resources Conservation Service of natural resource characteristics on non-Federal land in the United States. The inventory captures data on land cover and land use, prime farmland soils, soil erosion on cropland, wetlands, habitat types, and other resource-related attributes. We use these data to characterize the distribution of highly erodible cropland relative to the distribution of socially disadvantaged farmers.

REFERENCES

Ahearn, M. & Newton. D. (2009). *Beginning Farmers and Ranchers*, EIB-53, U.S. Department of Agriculture, *Economic Research Service*, 32, May.

Berry, J. K., Delgado, J. A., Khosla, R. & Pierce, F. J. (2003). "Precision conservation for environmental sustainability," *Journal of Soil and Water Conservation, 58(5)*, 332-39.

Brown, A., Jr., Christy, R. D. & Gebremedhin, T. G. (1994). "Structural Changes in U.S. Agriculture: Implications for African American Farmers," *The Review of Black Political Economy, 22(4)*, 51, Spring.

Caswell, M., Fuglie, K., Ingram, C., Jans, S. & Kascak, C. (2001). *Adoption of Agricultural Production Practices: Lessons Learned from the U.S. Department of Agriculture Area Studies Project*, AER-792, U.S. Department of Agriculture, Economic Research Service.

Cattaneo, A., Claassen, R., Johansson, R. & Weinberg, M. (2005). *Flexible Conservation Measures on Working Land: What Challenges Lie Ahead?* ERR-5, U.S. Department of Agriculture, *Economic Research Service.*

Claassen, R., Aillery, M. & Nickerson, C. (2007). *Integrating Commodity and Conservation Programs: Design Options and Outcomes,* ERR-44, U.S. Department of Agriculture, Economic Research Service.

Dismukes, R., Harwood, J. L. & Bentley, S. E. (1997a). *Characteristics and Risk Management Needs of Limited-Resource and Socially Disadvantaged Farmers,* AIB-733, U.S. Department of Agriculture, *Economic Research Service*, April.

Dismukes, R., Harwood, J. L. & Hoppe, R. A. (1997b). "Limited-Resource Farmers: Their Risk Management Needs," *Agricultural Outlook*, U.S. Department of Agriculture, Economic Research Service, May.

Feng, H., Kling, C., Glassman, P., Jha, M. & Parcel, J. (2006). "The Costs and Benefits of Conservation Practices in Iowa," *Iowa Agricutural Review, 12(4)*, 10-13, Fall.

Feng, H., Kling, C., Kurkalova, L., Secchi, S. & Gassman, P. (2005). "The Conservation Reserve Program in the Presence of a Working Land Alternative: Implications for Environmental Quality, Program Participation, and Income Transfer," *American Journal of Agricultural Economics, 87(5)*, 123 1-38.

Gundersen, C. (2008). "Measuring the extent, depth, and severity of food insecurity: an application to American Indians in the USA," *Journal of Population Economics, 21*, 191-215.

Hellerstein, D. (2006). "USDA Land Retirement Programs," *Agricultural Resources and Environmental Indicators, 2006 Edition*, K. Wiebe and N. Gollehon (eds.), EIB-16, U.S. Department of Agriculture, *Economic Research Service*, 175-83.

Hoppe, R., Korb, P., O'Donoghue, E. & Banker, D. (2007). *Structure and Finances of U.S. Farms*, EIB-24, U.S. Department of Agriculture, *Economic Research Service*,.

Jolliffe, J. (2004). *Rural Poverty at a Glance*, Rural Development Research Report 100, U.S.

Department of Agriculture, *Economic Research Service* No., http://www.ers.usda .gov/Publications/RDRR100.

Kott, P. (2001). "The Delete-a-Group Jackknife," *Journal of Official Statistics, 17(4)*, 521.

Lambert, D., Sullivan, P. & Claassen, R. (2007). "Working Farm Participation and Acreage Enrollment in the Conservation Reserve Program," *Journal of Agricultural and Applied Economics, 39(1)*, 151-69.

Lambert, D., Sullivan, P., Claassen, R. & Foreman, L. (2006). *Conservation- Compatible Practices and Programs: Who Participates?* ERR-14, U.S. Department of Agriculture, Economic Research Service.

Lerch, R. N., Kitchen, N. R., Kremer, R. J., Donald, W. W., Alberts, E. E., Sadler, E. J., Sudduth, K. A., Myers, D. B. & Ghidey, F. (2005). "Development of a conservation-oriented precision agriculture system: Water and soil quality assessment," *Journal of Soil and Water Conservation, 60(6)*, 411-20.

National Atlas of the United States. *Federal and Indian Land Areas*, Redlands, CA: ESRI, 2006, http://nationalatlas.gov/atlasftp.html?openCha pters=chpbound#chpbound.

Newton, D. & Ahearn, M. (2007). *Management Strategies for Beginning Farmers and Ranchers*, poster, American Agricultural Economics Association Annual Meeting, Portland, OR, http://purl.umn.edu/7629.

U.S. (2003). Department of Agriculture. *Environmental Quality Incentives Program Benefit Cost Analysis, Final Report,* May 21.

U.S. Department of Agriculture, Farm Service Agency. *Conservation Reserve Program: Summary and Enrollment Statistics, fy2007*, 2008, http://www.fsa.usda.gov/Interne t/FSA_File/annual_consv_2007.pdf.

U.S. Department of Agriculture, Natural Resources Conservation Service. *FY-2007 Conservation Program Allocations to States by Program*, 2007, http://nrcs.usda.gov/prog rams/2007_allocations/FY2007program_allocations_by_states

U.S. Department of Agriculture, Natural Resources Conservation Service. *Soil Survey Geographic (SSURGO) Database,* http://soildatamart.nrcs. usda.gov.

U.S. Department of Agriculture. "Farm Bill 2002: Conservation Opportunities for American Indians and Alaskan Natives, Beginning Farmers and Ranchers, and Limited-Resource Farmers and Ranchers," Fact sheet, June 2007, http://www.nrcs.usda.gov/program s/farmbill/2002/ pdf/ConOpFS_0905.pdf.

U.S. Department of Agriculture. *Socially Disadvantaged Farmers and Ranchers: Overview*, last modified 2006, http://www.fsa.usda.gov/FSA/w ebapp?area=home&subject=prod& topic=sfl.

Ward, E., Jemal, A., Cokkinides, V., Singh, G. K., Cardinez, C., Ghafoor, A. & Thun. M. (2004). "Cancer disparities by race/ethnicity and socioeconomic status," *CA: A Cancer Journal for Clinicians, 54(2)*, 78-93.

Wu, J. (2004). "Slippage Effects in the Conservation Reserve Program," *American Journal of Agricultural Economics, 82(4)*, 979-92.

Wu. J., Adams, R. M., Kling, C. L. & Tanaka, K. (2004). "From microlevel decisions to landscape changes: an assessment of agricultural conservation policies," *American Journal of Agricultural Economics, 86(1)*, 26-41.

End Notes

[1] This comparison uses 2007 Census of Agriculture data, which identifies the principal operator's race and ethnicity, and 1997 National Resources Inventory data that identifies cropland by erodibility level. The ARMS data used in this chapter define socially disadvantaged farmers and ranchers, but the Census data do not allow us to identify these farmers in precisely the same way (see appendix, "Data Sources"). However, the geographic distribution of these farmers is comparable between Census and ARMS.

[2] The Economic Research Service developed a classification system that categorizes farms based on sales, major occupation of the primary operator, and farm ownership structure. For a detailed description, see Hoppe et al., 2007.

[3] Production value reflects the market value of what is produced in a given year, regardless of whether it is shared among multiple parties or put into inventory. Farms may have no production due to adverse weather conditions, disease, or other reasons (Hoppe et al., 2007).

[4] When women are included in the definition of socially disadvantaged farmers, the proportion of socially disadvantaged farmers reporting zero production value rises from 20 percent to 33 percent.

[5] The 2008 Farm Act requires planting in 4 of the 6 years prior to 2008. Land coming out of CRP is considered planted, unless it is in trees.

[6] Although we are not able to identify socially disadvantaged farmers in the EQIP contract data, we are able to determine which contracts are on or near Indian reservations. See box, "EQIP Participation on American Indian Reservations."

[7] Some farms have more than one EQIP contract. However, counting contracts instead of farms does not appear to bias the representation of beginning and limited-resource producers.

[8] It is possible that preferences for particular programs may be driven more by farm size than farmer type. Data on the size of EQIP participant farms are not available, precluding us from drawing conclusions about the impact of farm size.

[9] Geographic data on the location of farms with EQIP contracts, State-level estimates of farm size, and Soil Survey Geographic (SSURGO) data allow us to identify highly erodible soils in the "neighborhood" of EQIP contracts. Here, we define the neighborhood as the average farm size in the State using ARMS data, and we allow this average to vary by farmer type if targeted farmers have farms significantly larger or smaller than other farmer types in the State.

[10] All tests described here are statistically significant at a 99-percent confidence level using a two-tailed t-test.

[11] CRP contract data contain information on the environmental problems that participants agree to treat in that program. The EBI score associated with each contract provides a relative ranking of the anticipated benefits (effectiveness of retiring land and installing contracted practices) in treating resource problems (see box, "Major Federal Conservation Programs"). Thus, scores vary by participant, depending on the physical characteristics of the participants' land and the effectiveness of practices that participants are willing to adopt. Although these data do not distinguish participants by farmer type and we cannot precisely identify the resource issues targeted farmers agree to treat in land retirement programs, the EBI allows us to characterize average expected benefits at a county level. For this analysis, we focus on variation in average scores for two factors that are part of the contract's EBI: the scores associated with soil erosion reduction and water quality benefits.

[12] The difference between farmer types' concentration in high-EBI-score counties was statistically tested using a binary Pearson's chi-square test statistic. The frequency of targeted farmers in counties with the highest 10 percent, 20 percent, and 30 percent of soil erosion and water quality EBI scores was significantly different from other farmer types at a 99-percent confidence level.

[13] Payments in land retirement programs include rental payments (CRP), easement payments (WRP), and cost-share payments for practice installation (CRP and WRP). Payments in CRP also include the Conservation Reserve Enhancement Program (CREP). The ARMS data on conservation program payments do not readily distinguish between the three payment types. This finding is consistent with Ahearn and Newton (2009).

[14] The 2001ARMS questionnaire included questions about types of conservation practices applied. These data allow some analysis of adoption patterns of CRP-eligible practices funded through land retirement programs, although the small sample sizes limit the reliability of tests for statistically significant differences between farmer types. Sample size limitations also preclude analyzing each of the three types of underenrolled farmers separately.

[15] To measure impacts on program efficiency, data on the opportunity costs of enrollment would be necessary.

[16] The ARMS survey asks the year that the operator began farming any operation. In 2006, this question was asked for up to three operators in only one version of the survey (version 1); previous years included the question on all versions of the survey. Counts of beginning farmer operations are potentially overestimated in 2006 if some farms with a less experienced primary operator have second or third operators that have more than 10 years of farming experience. Analysis of 2004 and 2005 data, however, show that frequency counts do not differ significantly if only the primary operator's farming experience is considered.

In: U.S. Farms and Conservation Programs
Editor: Lucas A. Savarese

ISBN: 978-1-61209-927-9
© 2011 Nova Science Publishers, Inc.

Chapter 3

THE FARM ACT'S REGIONAL EQUITY PROVISION: IMPACTS ON CONSERVATION PROGRAM OUTCOMES

Cynthia Nickerson, Marc Ribaudo and Nathaniel Higgins

ABSTRACT

The 2002 and 2008 Farm Acts increased funding for conservation programs that provide financial assistance to farmers to implement conservation practices on working farmland. Along with seeking cost-effective environmental benefits, these programs have a goal of spreading conservation funding equitably across States. The 2002 and 2008 Farm Acts strengthened this allocative goal by setting a minimum threshold for conservation funding for each State—one that exceeds historical funding for some States—for enrolling agricultural producers in specified conservation programs. This study uses conservation program data to examine evidence of the impacts of the Regional Equity provision of the 2002 Farm Act, and explores the tradeoffs that can occur among conservation program goals when legislation gives primacy to fund allocation. The study found that cross-State shifts in funding reduced the acres receiving conservation treatment for many resource problems, but increased the net economic benefits from treatments on some of them. Overall impacts on the types of producers enrolled were small.

ACKNOWLEDGMENTS

The authors thank David Nulph of USDA's Economic Research Service for assistance with figures using geographic information systems data. We also thank the following for their helpful comments on earlier drafts: LeRoy Hansen, Patrick Sullivan, and Marca Weinberg, all from ERS; Noel Gollehon from USDA's Natural Resources Conservation Service; Otto Doering from Purdue University; and two anonymous reviewers. Thanks also to Courtney Knauth for editorial assistance and Cynthia A. Ray for layout and design.

Summary

Federal working-land and farm and grazing land protection programs have long allocated program funding among States. In 2002, the Regional Equity provision of the Farm Security and Rural Investment Act (the 2002 Farm Act) further emphasized the allocative goal by redirecting some funding to States that historically had received only limited assistance under these programs. The Regional Equity provision mandated that each State receive at least $12 million per year between 2003 and 2007 for enrolling eligible producers in four conservation programs (the 2008 Farm Act increased the threshold to $15 million). This provision may sometimes work at cross purposes to the programs' environmental and economic goals of helping farmers adopt cost-effective conservation practices.

What is the Issue?

Because overall program budgets were not augmented to cover the minimum allocation requirement, the provision shifted funding from States that exceeded the threshold to those below it. The provision clearly affected the amount of conservation funding allocated to each State in each program. However, because conservation programs have multiple goals, it is difficult to predict the effects that promoting allocative goals may have on progress toward other program goals. Key questions are whether agricultural producers enrolled as a result of such legislation treat similar environmental problems and whether their conservation actions provide environmental benefits as cost effectively as actions undertaken by other producers, and also if particular types of farmers stand to gain or lose. This study analyzes program data from the 2004-06 period when the 2002 Farm Act was in effect. Data from the Environmental Quality Incentives Program (EQIP) and the Wildlife Habitat Incentives Program (WHIP)—the two programs with the most detailed data—are used to determine whether the required reallocations of funds affected the ability of those conservation programs to achieve environmental goals cost effectively.

What Did the Study Find?

The analysis revealed that the Regional Equity provision's $12 million threshold requirement had unequal impacts across States. Also, it altered the environmental and economic outcomes differently in EQIP and WHIP, in ways that were not always consistent across years. Major impacts evident from the study were:

- **The Regional Equity provision reduced the number of acres that received treatment for many resource problems in EQIP, but this did not always result in a decline in net economic benefits.** Changes in physical measures (such as acres receiving conservation treatment) are not always correlated with economic measures (the net economic benefits from conservation treatment), so it is important to measure both types of impacts in determining the tradeoffs arising from new policies. For instance, in EQIP, even though the cross-State shifts in enrollment induced by the provision decreased the number of acres treated for soil productivity and other soil erosion issues, in some years these cross-State shifts generated additional net economic benefits from improved soil productivity and reduced sedimentation (net economic benefits are calculated as gross benefits from conservation treatment minus treatment costs). This occurred when conservation actions benefited more people and when the producers in States that received increased funding (in order to reach the $12 million minimum) provided these benefits more cost effectively than producers in the remaining States. However, for grazing productivity and water conservation issues, the Regional Equity provision reduced both the number of acres receiving treatment and the corresponding net economic benefits from treatment in each of the 3 years studied.

- **Impacts differed among programs subject to the Regional Equity requirements.** In WHIP, the funding reallocation resulted in relatively large losses of both acres treated and net wildlife-related benefits (at least for the impacts the authors were able to measure in the continental United States), suggesting the provision may be having an overall negative impact on that program. Because each program targets different environmental problems at different costs, reducing the reallocations that occur in WHIP as a result of the Regional Equity provision and increasing reallocations that occur in the other three programs might increase overall net economic benefits from the four programs subject to the provision.
- **The 2002 Regional Equity provision had only a small impact on participation by types of producers that are offered more favorable enrollment terms in EQIP to encourage their involvement.** The study found that the Regional Equity provision reduced the number of livestock producers and increased the number of beginning and limited- resource producers enrolled in EQIP. However, the decrease in livestock producers was small and did not affect EQIP's ability to meet a legislated requirement that 60 percent of program funding be devoted to livestock- related practices.
- **The Regional Equity provision's fixed minimum funding threshold means that any decreases in total program budgets will be borne largely by States that exceed that threshold.** Because the Regional Equity provision is designed as a fixed set-aside and requires that each State receive at least the threshold amount of funding, any cuts in total program funds must be absorbed by States that receive more than the minimum funding. In contrast, program budget reductions would be shared by all States if the Regional Equity provision instead required that each State receive at least a certain percentage of total program funding. In the latter case, as total program budgets change, the funding for each State would change in proportion to the State's specified funding share.

How Was the Study Conducted?

This study uses conservation program contract data from 2004 to 2006 to identify the environmental, economic, and distributional implications of the Regional Equity provision within EQIP and WHIP, including the ability of the programs to deliver certain net benefits and the effect on enrollment patterns of certain producer groups over the period. Contract data were not available for the Farm and Ranch Lands Protection Program and the Grassland Reserve Program, the other two conservation programs subject to the Regional Equity provision. The analysis recognizes that the lowest ranked ("marginally accepted") contracts are the most vulnerable to policy- induced budget shifts. The authors used data on these contracts, including the costs incurred, acres treated, and resource concerns addressed—along with estimates of technical assistance costs and spatially heterogeneous data on the benefits of treating environmental problems arising from agricultural production—to estimate the impacts of the Regional Equity provision. Identifying the marginal contracts was particularly important because the lowest ranked contracts were more likely to address lower priority problems at higher cost. Due to limited data on the benefits of conservation treatment, not all conservation activity could be included in the analysis of economic impacts; our findings are illustrative of the tradeoffs that occur from implementing the Regional Equity provision. Also, the estimates of net benefits were limited to program activity in the continental United States.

INTRODUCTION

Over the past 20 years, most Federal assistance for conservation activities on agricultural land has been directed toward taking environmentally sensitive farmland out of production and enrolling it in the Conservation Reserve Program (CRP). Increasingly, however, Federal programs have sought to improve the environmental performance of active farms. Programs such as the Environmental Quality Incentives Program (EQIP) and Wildlife Habitat Protection Program (WHIP) have provided financial assistance to producers to implement conservation practices and adopt more environmentally friendly management practices. The Farm and Ranch Lands Protection Program (FRPP) and the Grassland Reserve Program (GRP) provide funds to permanently restrict conversion of farmland and grazing land to more environmentally damaging uses and, in the case of GRP, to restore native grasslands.[1] The 2002 Farm Security and Rural Investment Act (referred to here as the 2002 Farm Act) increased the combined spending for these four programs almost 400 percent to about $5.7 billion over the 6 years from 2002 to 2007. The Food, Conservation, and Energy Act of 2008 (hereafter called the 2008 Farm Act) authorized further funding increases, projected by the Congressional Budget Office to be almost $9.0 billion over the 5-year period from 2008 to 2012.[2]

The increase in funding for conservation on working lands has resulted in farmers' receiving more conservation assistance through programs that, by design, allocate program funds broadly across the country. USDA's Natural Resources Conservation Service (NRC S), the agency that administers the working-land conservation programs, allocates program funding to State NRCS offices based on conservation and other criteria. These State offices set priorities regarding which environmental problems will receive treatment and make enrollment decisions. Many State offices allocate funding and program decisions to sub-State (county, township, or watershed) offices. In contrast, in the CRP general signups in which most enrollments occur, applications are pooled nationally and enrollment decisions are made at the Federal level. Although some CRP funds are targeted to particular locations, most CRP program funds are awarded on the basis of expected environmental benefit and cost criteria that do not give particular consideration to location.

A decentralized program structure that furthers allocative goals can also help achieve environmental and economic goals. For example, when environmental problems arising from agricultural production are localized and the benefits and costs of treatment vary across the area, decentralized decisionmaking can benefit local jurisdictions economically because it lets them set priorities and determine solutions based on local needs and preferences (Peltzman and Tideman, 1972). Local decisionmaking would not necessarily advance the distributional goals of some conservation programs, however, such as ensuring that producers with different income levels and farming focuses have the opportunity to participate. Distributional goals, which target funds to particular types of farmers based on production characteristics, are distinguished from allocative goals, which are defined on the basis of geography.[3]

Recent legislative actions have furthered allocative goals in conservation programs. In the 2002 and 2008 Farm Acts, the Regional Equity provisions have required that each State receive a minimum amount of funding through, collectively, EQIP, WHIP, FRPP, and GRP.[4] The 2002 Farm Act set the minimum funding threshold at $12 million, and the 2008 Farm Act increased it to $15 million. The provision did not include an allowance to increase annual program budgets to cover these additional payments to States that historically had received lower levels of funding. Those States that had been receiving a higher amount continue to receive it if overall program funds permit; to ensure, however, that all States receive at least the required minimum annual funding, USDA reallocates conservation funds, when necessary, among States in the affected programs.

The Regional Equity provision affects multiple programs, so implementing it may involve a broader set of tradeoffs with other program goals than requirements that affect a single program. Each year since it was first passed in 2002, the provision has resulted in reallocation of funds, in each of the four targeted programs, from States that exceeded the minimum threshold to those that did not. This provision allows additional lower priority applications to be enrolled in States that are allocated more funds and prevents lower priority applications from being enrolled in the other States. Funding reallocations across States can affect program outcomes in ways that are difficult to predict because a program's applications that are viewed as low priority in one location could be viewed as higher priority, or not a priority at all, in another location. Program outcomes could change because the characteristics of the applications—and the environmental problems the operators agree to treat— differ across areas (that is, spatially) and across conservation programs. Agricultural operations across the United States are diverse, with both the types of production and characteristics of farms varying geographically. The severity of environmental problems arising from agricultural problems also varies geographically, as does the willingness of operators to implement conservation practices to address them (Lambert et al., 2006; Caswell et al., 2001; Claassen et al., 2001). Local government priorities in each conservation program vary geographically as well. Collectively, these variations mean that operators in different locations will face different incentives to enroll in conservation programs.

If programs initially were strictly designed to enroll producers who could provide the most environmental benefits for the least cost, we might expect the tradeoff from expanding participation in selected geographic areas or for particular producers to reduce the cost effectiveness of conservation programs. Cost effectiveness can decline when program changes result in fewer benefits for a given expenditure or in the same level of benefits at a higher cost. However, policy changes to strengthen allocative goals may not necessarily occur at the expense of economic goals. Conservation programs are not designed to single-mindedly maximize environmental benefits, but instead attempt to satisfy multiple goals. As a result, reallocating program funds can have impacts that are hard to predict. Whether a policy change entails a tradeoff between the cost-effective delivery of environmental benefits and other program goals depends on the relative values of benefits and costs provided by the contracts that ultimately are, and are not, funded as a result of the legislated change.

The literature provides little guidance on this question. Two recent ERS analyses found that efforts to further distributional goals may have increased conservation program costs. One study considered the impact of discontinuing competitive bidding in EQIP—the option to bid down payments to improve chances of enrollment (Cattaneo et al., 2005). Bidding was discontinued in 2002 in EQIP because it was perceived as favoring large and well-established producers over small and limited-resource farmers. This mandated change affected program costs: the average payment for implementing structural conservation practices was 35 percent of cost when bidding was allowed (1996-2002), but after bidding was discontinued, producers received the minimum Government cost-share of at least 50 percent, and many producers received the maximum cost-share of 75 percent (Cattaneo et al., 2005).[5] ERS analysis of post-2002 enrollment patterns revealed that they did not shift appreciably to States that—by some estimates—may be likely to generate higher levels of environmental benefits (Hansen and Ribaudo, 2008), which suggests that discontinuation of bidding may have reduced EQIP's ability to provide benefits in the most cost-effective manner (an economic goal). Another study found that because beginning and limited-resource farmers enrolled in EQIP receive favorable payment terms, they tend to receive larger payments per practice, even though they tend to treat fewer acres or install smaller structural practices. But they also operate more environmentally sensitive land than other participating farmers, have different conservation priorities, and receive different levels of payments (Nickerson and Hand, 2009). The differences among farmers who currently participate suggest that economic and environmental outcomes could change if more of these farmers were enrolled.

The Regional Equity provision furthered allocative goals, described in the following section, that already existed in the working-land programs. To determine how the provision's stipulations for allocating funds affected program results, the authors examined data that help identify impacts on key program outcomes. This chapter describes the design structure of working-land conservation programs that allows them to accommodate allocative goals, and analyzes the impacts on conservation program outcomes of the Regional Equity provision that further these goals. The authors examine several key issues. First, we analyze the effects of the Regional Equity-induced funding reallocations on the number and types of contracts that were enrolled. We then evaluate the tradeoffs that occurred in terms of the resource problems that were ultimately treated. We use these findings to shed light on whether the Regional Equity provision reduced the ability of working-land programs as a group to achieve environmental and economic goals—by providing environmental benefits cost effectively—over the 2004-06 period. We also consider the impact of the Regional Equity provision on distributional goals by considering its effects on enrollments of livestock producers and beginning and limited-resource producers in EQIP—groups that are given favorable enrollment terms to encourage their participation. We use a methodological approach that identifies characteristics of contracts most susceptible to such changes, which is particularly important in decentralized programs that allow for priority setting and enrollment decisions at a local level. Policymakers have a limited set of levers for affecting change in conservation programs, which increases the importance of understanding the potential tradeoffs from changes in programs that seek multiple goals simultaneously.

ALLOCATIVE GOALS IN WORKING-LAND AND LAND PROTECTION PROGRAMS

Along with environmental and economic goals of providing benefits cost effectively, regional funding allocation considerations have historically played an important role in the design of working-land and farm and grazing land protection programs. Since the inception of the programs, financial and technical assistance funds to address a variety of environmental problems have been made available to producers in all 50 States through EQIP, WHIP, FRPP and GRP (figure1).

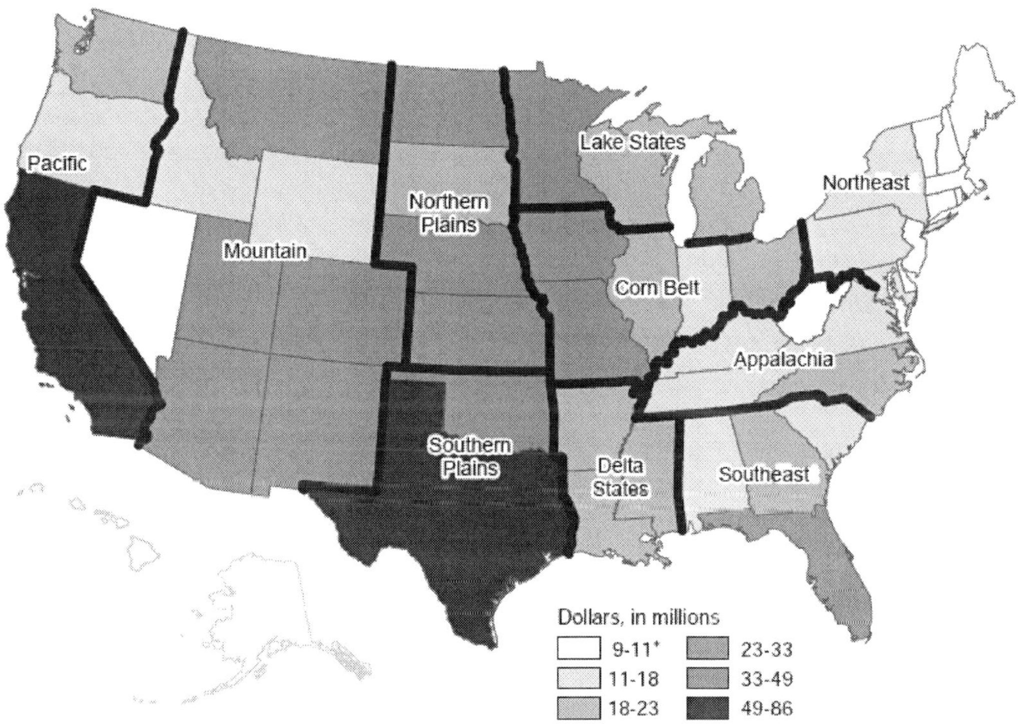

EQIP - Environmental Quality Incentives Program; WHIP = Wildlife Habitat Incentives Program; GRP Grassland Reserve Program; FRPP = Farm and Ranch Lands Protection Program.

*In 2006 only, NRCS administratively lowered the Regional Equity threshold from $12 to $11 million.

Source: ERS analysis of NRCS fund allocation data. Map excludes funds allocated to Puerto Rico and the Pacific Basin.

Figure 1. Initial funding for EQIP, WHIP, GRP, FRPP, 2006 (pre-Regional Equity reallocations).

In EQIP, WHIP, FRPP, and GRP, determining the funding each State NRCS office will receive involves two steps. First, in each program, most funds available for financial assistance are allocated using indicators reflecting national priorities in farm legislation.[6,7] The specific types of indicators vary across programs (Appendix A). NRCS weights each indicator based on an assessment of the relative importance of the corresponding resource issue. Indicator weights are adjusted from time to time as new information or issues come to light. Second, funds for each indicator are then allocated to State NRCS offices based on the extent of the environmental concern that the indicator represents in each State. For example, if land condition is a program indicator, States with higher percentages of poor quality land will receive a larger percentage of the funds allocated to the land condition indicator.

In each program, the total allocation of financial assistance funds for each State is the sum of the State allocations for each of the indicators. Allocating funds to States based on the *relative* extent and impairment of resources, rather than on *absolute* quantities, ensures that at least some funds from each program will be allocated to each State NRCS office (at least, for those indicators that are present in the State).

When allocative goals are sought, conservation funds in working-land and farm and grazing land protection programs tend to be less geographically concentrated among States than they would be otherwise. For example, in the Conservation Reserve Program, in which broad distribution of funds is not an explicit element of program design (applications are pooled nationally and prioritized on the basis of expected environmental benefits and costs), more than 56 percent of program outlays were concentrated in eight States in 2006. Each of these eight States individually accounted for more than 5 percent of outlays (figure 2).[8] In contrast, in EQIP, a decentralized program, only one State was allocated more than 5 percent of program funding in 2006. The eight States receiving the largest allocations in EQIP accounted for 33 percent of total funding that year.

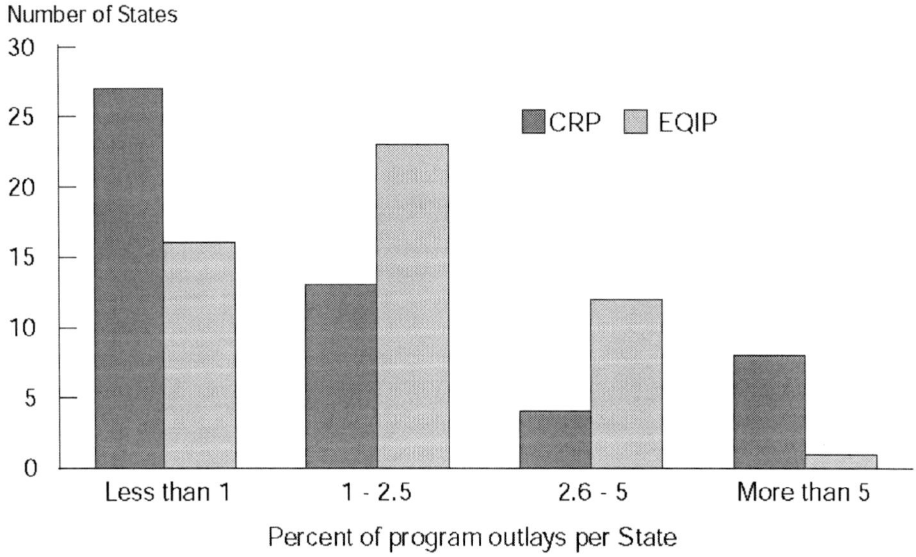

CRP = Conservation Reserve Program; EQIP = Environmental Quality Incentives Program.
Sources: USDA-FSA CRP summary data; USDA-NRCS EQIP summary data.

Figure 2. Comparative distribution of program funding to States, 2006.

Even when conservation funding is allocated across States, program outcomes might not change substantially if all States had the same environmental priorities and agreed to fund the same conservation practices at the same cost. But State-level priorities and decisions differ. These differences have an important bearing on which producers choose to apply for funding, as well as which are enrolled and receive conservation payments. The discretion given to local NRCS offices in these programs introduces variations that make the impacts of national legislation on program outcomes harder to predict.

THE REGIONAL EQUITY PROVISION: DID IMPLEMENTATION INDUCE ENVIRONMENTAL, ECONOMIC, OR DISTRIBUTIONAL TRADEOFFS?

An analysis of the Farm Act's Regional Equity provision reveals how tradeoffs can arise from furthering allocative goals in conservation programs. Although minimum funding thresholds were set in the 2002 and 2008 Farm Acts at $12 million and $15 million, respectively, the Regional Equity provision did not dictate how the required minimum thresholds were to be met. During the 2004-06 period that we analyzed, USDA's Natural Resources Conservation Service (NRCS) determined how funds for the four programs would be reallocated to meet the requirement and the programs in which reallocations would occur.[9] During this period, only in 2005 did reallocations occur in all four affected programs (table 1). In 2004, NRCS reallocated funds within EQIP, GRP, and FRPP to meet the requirement. In 2006, funding was not appropriated for GRP, so reallocations occurred only within EQIP, WHIP, and FRPP.

The reallocation that occurred within each program also varied each year. In EQIP, declining funds (in both total dollars and percentage of funding) were set aside specifically for States below the funding threshold over the 2004-06 period, while in the other three programs an increasing percentage of funds was set aside. Most of the reallocations occurred through EQIP, but in 2006, WHIP and FRPP were particularly impacted by the Regional Equity provision: 22 and 38 percent of the Regional Equity reallocations occurred through these programs, even though the programs represented only 4 and 8 percent of the combined program budgets, respectively.[10]

Over the 2004-06 period, the Regional Equity provision tended to shift program funds to 12 States—mostly the smaller States in the Northeast, but also to Nevada, Alaska, and Hawaii (figure 3). We define "Regional Equity States" (RE States) as the States that were initially below the minimum funding threshold and were then reallocated more conservation funds, and the "non-Regional Equity States" (non-RE States) as the remaining States.

State Funding Levels Will Not Always Be Equally Affected by Changes in Overall Program Budgets

With its minimum allocations based on a fixed amount, the Regional Equity Provision has broad implications for which States gain or lose funding when overall conservation program budgets change. In the absence of the Regional Equity provision, declines or increases in overall program budgets would generally be shared equally by all States in proportion to their funding share. In the presence of the Regional Equity provision, however, overall program budget reductions will be absorbed by the majority of States whose allocated funding exceeds the minimum threshold amount, since States with funding levels at the fixed threshold are required to receive at least the threshold amount. As program budgets rise, States whose funding exceeds the threshold share in the increase, and fewer reallocations are necessary to assure that the other States are allocated the minimum required by the Regional Equity provision. If budgets become large enough that the Regional Equity provision no longer forces reallocation (all States' budgets exceed the threshold), all States will share in the budget increase in proportion to their funding shares.

These unequal impacts of budget changes under the Regional Equity provision's fixed set-asides contrast with the impacts of set-asides that advance distributional goals in EQIP. These other set-asides are specified as proportional amounts, such as the 60 percent of the program budget set aside for livestock practices, 5 percent for beginning farmers and ranchers, and 5 percent for socially disadvantaged farmers and ranchers. In these cases, the amounts ultimately reserved for targeted groups vary in proportion to program budget changes, so all groups are relatively equally affected by overall funding changes.

Note: "Regional Equity State" means State received funds from other States due to the Regional Equity requirement. South Carolina did not receive additional Wildlife Habitat Incentives Program (WHIP) funds in any year; Maine did not receive WHIP funds in 2006.

Source: ERS analysis of NRCS fund allocation data. Map excludes Puerto Rico, an RE 'State' in 2004-2006.

Figure 3. Regional Equity status.

Table 1. Funding levels and Regional Equity adjustments.

Fiscal Year	EQIP	WHIP	GRP	FRPP	Total
2004 Total initial Financial Assistance (FA) funding[1]	$661,005,400	$27,800,000	$54,059,400	$84,833,000	$827,697,800
Percent of total initial FA funding for all 4 programs	80%	3%	7%	10%	100%
Amount of FA funds reallocated to Regional Equity States	$31,476,600	$0	$4,489,700	$8,640,000	$44,606,300
Share of Regional Equity funds reallocated through program	71%	0%	10%	19%	100%
Percent of programs' initial FA funding affected by reallocations	5%	0%	8%	10%	
Number of Regional Equity States[2]	14	0	14	10	
Number of non-Regional Equity States[2]	38	0	38	42	
2005 Total initial FA funding[1]	$691,525,000	$33,050,000	$65,960,000	$106,700,000	$897,235,000
Percent of total initial funding for all 4 programs	77%	4%	7%	12%	100%
Amount of funds reallocated for Regional Equity - FA	$25,500,000	$6,240,000	$9,790,000	$12,000,000	$53,530,000
Share of Regional Equity funds reallocated through program	48%	12%	18%	22%	100%
Percent of programs' initial FA funding affected by reallocations	4%	19%	15%	11%	
Number of Regional Equity States[2]	14	13	13	12	
Number of non-Regional Equity States[2]	38	39	37	40	
2006 Total initial FA funding[1]	$697,100,000	$29,170,000	$0 0% $0 0% 0%	$66,040,000	$792,310,000
Percent of total initial funding for all 4 programs	88%	4%	0	8%	100%
Amount of funds reallocated for Regional Equity - FA	$18,700,000	$10,270,000	0	$17,460,000	$46,430,000
Share of Regional Equity funds reallocated through program	40%	22%		38%	100%
Percent of programs' initial FA funding affected by reallocations	3%	35%		26%	
Number of Regional Equity States[2]	13	12		12	
Number of non-Regional Equity States[2]	39	40		40	

EQIP=Environmental Quality Incentives Program; WHIP=Wildlife Habitat Protection Program; GRP=Grassland Reserve Program; FRPP=Farm and Ranch Lands Protection Program.

[1] Analysis is limited to initial funding allocations for financial assistance. Performance bonus funding is excluded.

[2] Regional Equity (non-Regional Equity) States are those that benefited (did not benefit) from the RE provision.

The number of RE and non-RE States includes Pacific Basin and Carribean. South Carolina became a non-RE State in EQIP in 2006. Maine became a non-RE State in WHIP, in 2006.

Source: NRCS data on Regional Equity allocations.

Participation Patterns Differ between RE and Non-RE States

Conservation program participation patterns differ between Regional Equity and non-Regional Equity States, suggesting that Regional Equity-induced funding shifts can impact program outcomes in ways that are hard to predict.

The programs subject to the Regional Equity provision depend on producers' volunteering to participate—to implement conservation practices, restore grassland, or sell an easement—in exchange for program payments. The requirements imposed by the Farm Act's Regional Equity provision result in some States' receiving additional conservation money to meet the minimum funding threshold even when producer interest in participating and demand for program funding in those States is not strong. For example, in 2004 in EQIP, to meet the Regional Equity requirements, funds were shifted away from non-RE States that had, on average, three times more demand for program funds than RE States. In other programs, however, the Regional Equity provision resulted in more funding being allocated to States where producer interest in participating was higher. In FRPP, funds were shifted *toward* RE States with more than twice the average demand (demand is measured in terms of the dollars associated with unfulfilled contract applications). States with excess demand by producers experience the greatest competition for program funds, and they are the places where programs may have the greatest opportunity to achieve a given level of benefits at the least possible cost.[11]

If little variation exists across States in the types of environmental problems treated and in the costs and benefits of the contracts funded within each program, satisfying an allocative objective by shifting program funds between RE and non-RE States will have little effect on program outcomes. However, several considerations suggest impacts of shifting funds are likely to be nontrivial. The four affected programs (EQIP, WHIP, FRPP, and GRP) are designed to address a variety of agriculturally related environmental issues, and the severity of these problems varies across the country. For example, water conservation has historically been a concern in the Western United States, where drought-related problems exist, and reducing fertilizer and nutrient runoff has been a concern throughout the country, especially in areas such as the Corn Belt and Chesapeake Bay watershed, where water quality problems impact local economies (see, for example, Bockstael et al., 1989, Leggett and Bockstael, 2000, and Secchi et al., 2007).

State NRCS offices have had considerable flexibility in working-land programs in determining which resource problems will receive priority in any given year, as well as flexibility with payment rates for practices that are cost-shared, and these priorities can vary even across neighboring States. Producers vary across States in their willingness to adopt particular practices and in terms of the costs they face even when adopting the same practice. The social benefits of conservation treatment can also vary across States; some types of benefits are highest when actions are concentrated in high- population States (Hansen and Ribaudo, 2008; Feather et al., 1999). Taken together, the variation in environmental problems, State-level priorities, producer preferences, and costs and benefits suggests that shifting funds from one State to another could result in very different conservation outcomes.

Summary statistics for two programs, EQIP and WHIP, suggest that by several measures—including costs, types of practices, and resource problems receiving treatment—a shift in program funds between RE States and non-RE States would result in different types of contracts being funded. In terms of costs, the Regional Equity provision has shifted funds to States with higher contract payments on average, although in both groups of States wide variation exists. [12] These average contract payments ranged from about $24,000 to $29,000 per contract in RE States over 2004-06, while in non-RE States they ranged from about $15,000 to $18,000 (table 2). In WHIP, RE States funded contracts that were twice as large, on average, due primarily to large contract payments in Alaska and Hawaii.

One reason contract payments were higher in RE States was that those States tended to reimburse participants for cost-shared practices at a higher rate than non-RE States. EQIP and WHIP provide cost-share assistance to producers who install structural practices (e.g., manure storage, irrigation systems, fencing, and field-edge vegetative buffers) and financial assistance to those adopting conservation-compatible management practices like nutrient management and prescribed grazing. In EQIP, the largest proportion of contracts in RE States received cost-share payments for structural practices at a 75-percent rate, while in non-RE States most contracts were cost-shared at a 50-percent rate. In WHIP, three-quarters of the contracts in RE States received a 75-percent cost-share rate compared with half of contracts in non-RE States that received that rate.

Table 2. Characteristics of Regional Equity and non-Regional Equity States.

		EQIP		WHIP	
	Fiscal year	**RE States**	**non-RE States**	**RE States**	**non-RE States**
		Dollars			
Average payments - all contracts	2004	$23,400	$15,000	-	-
	2005	$22,500	$15,300	$16,700	$8,500
	2006	$28,800	$17,800	$23,600	$8,900
Cost-share rates[1]		*Percent of Con tracts[1]*			
50% cost-share rate	2005	28%	52%	3%	18%
51 - 74% cost-share rate	2005	19%	30%	22%	28%
75% cost-share rate	2005	39%	13%	75%	54%
> 75% cost-share rate	2005	14%	5%	0%	0%
Practice type					
Structural only	2006	41%	51%	33%	46%
Management only	2006	31%	26%	48%	34%
Structural and management	2006	28%	24%	18%	19%
Spending by resource concern		*Percent of contract costs*			
Soil condition	2006	8%	8%	0%	0%
Soil erosion	2006	14%	17%	18%	9%
Domestic animals	2006	8%	11%	0%	1%
Plant condition	2006	17%	16%	6%	3%
Water quantity	2006	11%	13%	2%	2%
Water quality	2006	37%	29%	3%	2%
Air quality	2006	3%	3%	0%	0%
Fish and wildlife	2006	3%	3%	72%	84%

EQIP = Environmental Quality Incentives Program; WHIP = Wildlife Habitat Incentives Program.

[1]Some contracts contain multiple practices cost-shared at different rates, and for these contracts the (weighted) average cost-share rate falls between the standard 50%, 75%, or 90% rates.

Source: ERS analysis of NRCS contract data.

The higher average cost of contracts in RE States suggests that fewer contracts may have been funded across all States in EQIP and WHIP because of the Regional Equity provision.

However, higher costs do not neces-sarily mean that program benefits were reduced. RE States tended to fund contracts with producers who agreed to address more resource concerns and adopt more practices. Further, in EQIP, the RE States tended to fund more contracts that implemented combinations of practices—that is, more producers in RE States agreed to implement both structural and management practices to address environmental concerns. Implementing a group of practices, as in a conservation system, can be more effective at addressing environmental problems than single practices (e.g., Lerch et al., 2005; Berry et al., 2003).

The resources receiving priority, as measured by costs incurred for treating problems, differ somewhat between the RE and non-RE States. In 2006 in EQIP, the largest share of contract funds in both groups of States was devoted to treating surface water quality problems, with RE States spending 37 percent of funds on these issues compared with 29 percent in non-RE States (spending patterns in previous years were similar). In addressing other issues, the share of funds spent by both groups of States was more similar (table 2). Due to WHIP's narrower focus on species and habitat protection, it is not surprising that, in both groups of States, most WHIP funds are spent on fish and wildlife issues—72 percent and 84 percent of costs in RE and non-RE States, respectively—although different species may benefit in different States (table 2). Also, a greater share of contract costs in the RE States was incurred to treat soil erosion—18 percent compared with 9 percent in non-RE States.

The Regional Equity Provision Entailed Economic and Environmental Tradeoffs

The differences between conservation program contracts and the general types of practices adopted in the RE versus non-RE States are indicative of some of the likely impacts of the Regional Equity provision. However, it is the specific characteristics of the additionally enrolled contracts in RE States, and of the applications that did not receive funding in non-RE States, that ultimately determine the economic and environmental tradeoffs resulting from enactment of the Regional Equity provision. These characteristics also determine whether tradeoffs occurred between allocative and distributional objectives, such as whether certain producer types gained or lost conservation funding.

Data on specific contracts funded through EQIP and WHIP, coupled with an underst-anding of the fund allocation formulas used by the conservation programs, demonst-rate some of the environmental, economic, and distributional impacts of the Regional Equity provision over the 2004-06 period in those programs. Our analysis recognizes that the lowest ranked contracts ("marginal contracts") are the most vulnerable to the funding shifts resulting from Regional Equity reallocations. We estimated which marginal contracts were affected by the Regional Equity-induced changes in State budgets, and the characteristics of these marginal contracts formed the basis for our estimates of environmental and economic tradeoffs of the Regional Equity provision (see box, "Methods").

Using a methodology that relies on information from the marginal contracts to estimate the effects of provisions furthering an allocative goal has a number of advantages. By definition, the marginal contracts are those least preferred for enrollment, which means they are less likely to address the resource concerns of highest priority to the jurisdiction, or they do so at higher cost.

METHODS

To estimate the effects of the Regional Equity provision, we used the programs' Federal funding allocation formulas to approximate how States' funding for financial assistance changed due to the Regional Equity provision. In the RE States, the Regional Equity provision expanded State funding to bring them to the $12 million threshold, so the contracts affected by the provision were those funded with the additional money. In these States we identified the "marginal" contracts using the ranking score of each application, and assumed those applications ranked lowest were the additional contracts. In the non-RE States, the RE provision reduced State budgets, resulting in some applications' remaining unfunded. Because the Regional Equity provision was enacted during a period of increasing program budgets and high producer demand for conservation funding, we assumed that the changes in State-level budgets induced by the provision did not affect producers' incentives or decisions to apply to the conservation programs.

Ideally, data on proposed practices and the acres and resource issues to be treated would be available from a review of all conservation program applications, and these data could be used to characterize how outcomes would have changed if additional applications had been funded in the non-RE States. However, such data are available only on contracts that received funding. In the absence of applicant data, several alternatives exist for proxying the practices and actions applicants would be willing to undertake if additional funds were available to increase enrollments. One alternative is to assume that previously unfunded applicants' parcels have similar characteristics to the entire farmland population. Using simulations coupled with program cost data, we could estimate, on average, how many acres applicants would be willing to treat, for which resource concerns, and at what cost. However, this strategy may not be ideal in programs like EQIP and WHIP, which, our analyses show, enroll a non-homogenous set of contracts in any given year (see Appendix B). We used a different strategy, one that assumes an expanded budget would enroll land that most closely resembles a much narrower segment of farmland: the applications that got funded but were of the lowest priority for enrollment—those contracts that were ranked just slightly higher than rejected applications. Assuming similarity between these two groups implies that producers with low-priority contracts that were accepted and those that just missed being accepted have similar expectations about the profitability of implementing particular practices to mitigate particular physical effects.* By comparing the characteristics of marginal contracts in the RE States to those in the non-RE States, we were able to discern how acres receiving treatment, the types of physical effects receiving treatment, practice types and costs, and participant types were affected by the Regional Equity provision.

* This approach also assumes that a sufficient amount of acreage remains that needs treatment for the resource concerns that were addressed by the marginally accepted contracts. EQIP and WHIP have historically had a large backlog of applications, so it is plausible that an adequate supply of land needing treatment exists

An analysis of contract data reveals that the marginal contracts in States that benefited from the Regional Equity reallocations have different characteristics than nonmarginal contracts.

Our analysis involved three steps to identify environmental and economic impacts of the Regional Equity provision:

- First, to estimate environmental impacts, we used the data on marginal contracts affected by the Regional Equity provision to estimate the change in the number of acres that received treatment for various environmental problems as a result of the funding shifts across States.
- Second, to estimate economic impacts, we analyzed the costs of treating environmental problems associated with the contracts affected by the Regional Equity provision (see Appendix C). Because the goal of the economic analysis was to compare changes in treatment costs to changes in benefits from the treatment effort, the cost analysis was limited to the subset of conservation treatment activities for which benefit estimates were available. These were activities undertaken to reduce erosion from wind and water (to improve onsite soil productivity and water and air quality and reduce sedimentation); to improve grazing productivity, water conservation, wildlife habitat, and pheasant-hunting opportunities; and to create wetlands. EQIP and WHIP contracts address more issues than this limited set of activities, but a lack of information on the benefits of treating other issues meant a substantial portion of conservation activity was excluded from our analysis. In EQIP, 24-50 percent of contract costs mapped to these nine benefit categeories over the study period; in WHIP, 56-75 percent did so (see box, "Conservation Activities Linked to Environmental Benefits").
- Finally, for each category of conservation treatment we considered, we combined per acre cost and benefit data with estimates of the number of acres treated in the watershed in 2004-06 to find out whether the changes in conservation treatment resulting from the Regional Equity provision resulted in additional net benefits or costs (see Appendix D for a description of the studies from which the earlier benefit estimates were derived).

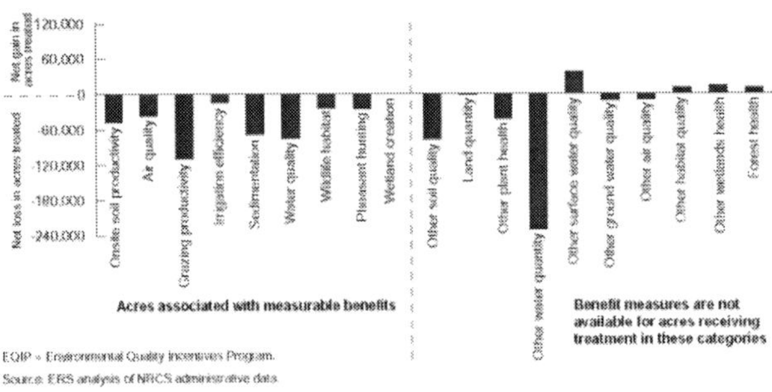

EQIP = Environmental Quality Incentives Program.
Source: ERS analysis of NRCS administrative data.

Figure 4. EQIP 2004: estimated net gain/loss in acres treated due to Regional Equity provision.

CONSERVATION ACTIVITIES LINKED TO ENVIRONMENTAL BENEFITS

Conservation programs provide a wide range of environmental and social benefits. While important, many of these benefits are difficult to quantify in monetary terms that enable a comparison to program costs. Due to limited data on the benefits of conservation treatment, a relatively small proportion of contract costs for EQIP were included in the economic analysis of Regional Equity impacts. In 2004 and 2005, only 24 percent of the costs of marginal contracts—those contracts affected by the Regional Equity provision—were incurred to treat problems for which benefit estimates were available (see the EQIP contract costs figure below). In 2006, 50 percent of contract costs were related to the nine benefit categories. In WHIP in 2005, 56 percent of marginal contract costs were related to the seven benefit categories addressed by that program, compared with 75 percent in 2006 (see the figure). There could be a number of reasons for the difference in the amount of conservation activity associated with measurable benefits in 2006. Local government priorities can change yearly, as can the financial incentives producers have to address particular resource problems. We found reasonable consistency across years between conservation practices and the resource problems that local field offices indicated were associated with particular resource concerns (see Appendix C). In all years, the analyses were limited to program activity in the contiguous 48 States. Because benefit estimates are not available for any resource issues in Alaska and Hawaii or the territories, all costs associated with contracts funded in these places were excluded from our analysis.

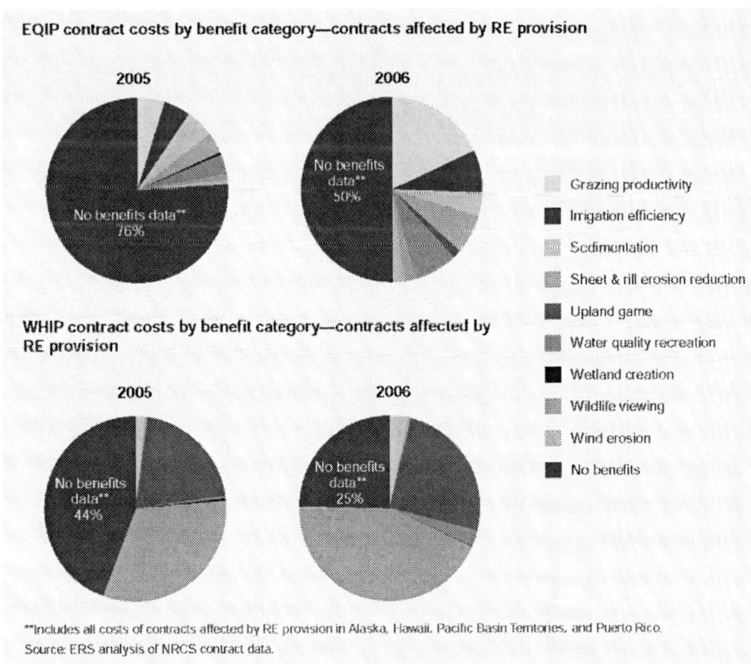

Using more aggregated data, such as the characteristics of the "average" contract—which is tempting because it would involve simpler calculations—can be misleading for policy analysis (see Appendix B). Our method also takes into account the variation in local priorities afforded by the decentralized structure of working-land programs. Other methodological approaches are less able to accommodate this variation, which is likely to have important impacts on local enrollment incentives and program outcomes.

The Regional Equity Provision Affected the Number of Acres Treated

Estimates of the acres receiving treatment reveal that the impacts of the Regional Equity provision differed over the study period (see Appendix C for methods used to estimate acres subject to treatment). In 2004 and 2005, the shifts in funding from the non-RE States to the RE States resulted in net losses in the acres receiving conservation treatment for a number of environmental problems in EQIP and, in 2005 and 2006, also for WHIP. For example, in EQIP in 2004, over 70,000 fewer acres received treatment for grazing productivity, sedimentation, and water quality problems and over 200,000 fewer acres were enrolled for water quantity issues (figure 4).[13] In 2005, except for reductions in the amount of land receiving treatment for grazing productivity issues, the net changes in acres receiving treatment were not as dramatic in EQIP (figure 5). In 2006, however, the Regional Equity reallocations resulted in significant net gains in treated acres for plant and soil condition issues in EQIP (figure 6).

In WHIP, program funding levels were significantly lower than in EQIP, and fewer resource concerns were treated. The biggest impacts were net losses of 20,000 to 55,000 acres treated each year for pheasant-hunting and wildlife habitat (figures 7 and 8).

Costs and Benefits Vary between Contracts Funded in RE States versus Contract Applications in non-RE States

An analysis of per acre treatment costs for the selected resource concerns revealed that contracts funded with Regional Equity funds in RE States had higher per acre costs than the contracts that would have been funded, in the absence of Regional Equity restrictions, in non-RE States (table 3).[14] This finding held throughout the 3-year study period, in both EQIP and WHIP, almost without exception.[15] Costs varied over a wide range, however. In some cases, the maximum cost per acre estimate exceeded the mean per acre estimate by several thousand dollars in the RE States, due to the installation of some expensive practices (e.g., micro-irrigation systems, waste storage facilities, and ponds), along with implementation of a broad combination of structural, vegetative, and management practices. In some cases, Regional Equity funds were used to install less typical practices at a high cost per acre, such as shellfish aquaculture management systems.

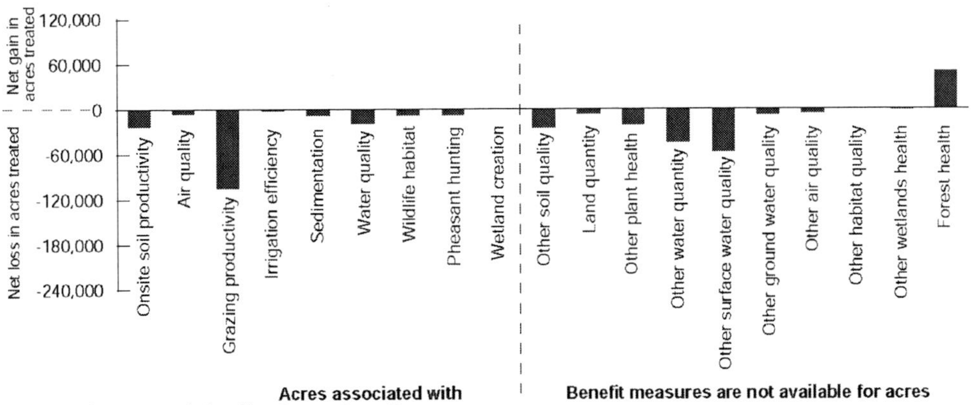

EQIP = Environmental Quality Incentives Program.
Source: ERS analysis of NRCS administrative data.

Figure 5. EQIP 2005: estimated net gain/loss in acres treated due to Regional Equity provision.

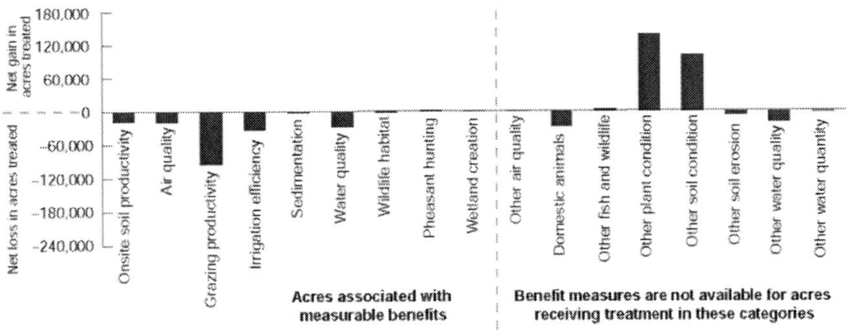

EQIP = Environmental Quality Incentives Program.
Source: ERS analysis of NRCS administrative data.

Figure 6. EQIP 2006: estimated net gain/loss in acres treated due to Regional Equity provision.

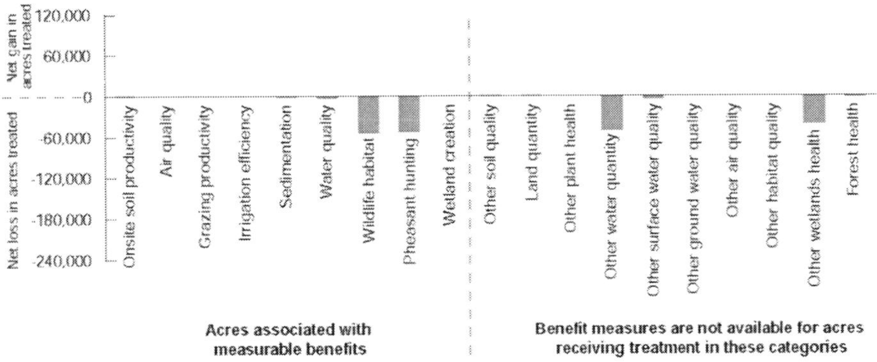

WHIP = Wildlife Habitat Incentives Program.
Source: ERS analysis of NRCS administrative data.

Figure 7. WHIP 2005: estimated net gain/loss in acres treated due to Regional Equity provision.

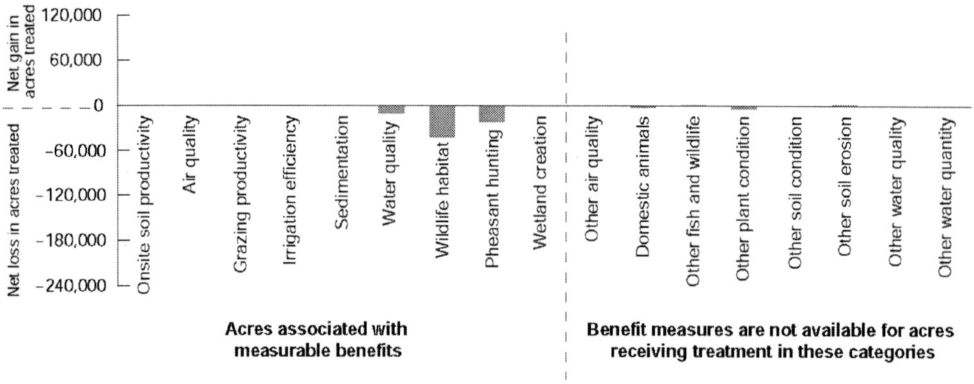

WHIP = Wildlife Habitat Incentives Program.
Source: ERS analysis of NRCS administrative data.

Figure 8. WHIP 2006: estimated net gain/loss in acres treated due to Regional Equity provision.

The benefits from conservation treatment also appeared to vary between RE and non-RE States, at least for the subset of conservation activities for which benefit estimates were available. Regionally varying estimates of benefits that were drawn from a number of studies reveal that the average per acre benefits from reducing sedimentation and increasing water quality were higher in RE than non-RE States (table 4). (See Appendix D for a description of the studies from which the benefit estimates were derived.) This suggests that shifting conservation efforts to RE States to mitigate these problems could produce greater conservation benefits. The benefits from creating wetlands are an order of magnitude higher than the other benefits, because wetlands generate a whole set of benefits, including improved water quality, wildlife habitat, and recreational opportunities. Though wetland-related benefits are likely to vary by area, regional estimates were not available.

Despite reducing the number of acres that received treatment for many resource problems, the RE provision increased the net benefits from treating some of them (net benefits are calculated as gross benefits minus costs). The RE provision allowed EQIP to generate an estimated additional \$4.5 million and \$2.4 million in sedimentation reduction benefits in 2005 and 2006, and an estimated additional \$490,000 in soil productivity benefits in 2004 and 2005 (figure 9). These changes occurred because the acres that were treated in RE States (in those treatment areas for which we have benefit measures) generated higher levels of benefits than the acres that would have received treatment in the non-RE States, due partly to higher surrounding population levels that benefit from improved environmental conditions. In terms of grazing productivity benefits, the RE provision generated net losses of about \$2 million yearly. The effects also varied by year. For the nine categories of environmental issues considered, the RE provision collectively resulted in losses of net benefits in EQIP of about \$2.5 million in 2004, a gain of \$2.3 million in net benefits in 2005, and losses of \$3 million in 2006.

Table 3. EQIP contract costs per acre by benefit type - 'marginal' contracts affected by the Regional Equity provision, 2006[1].

	Onsite soil productivity benefits	Air quality benefits	Grazing productivity benefits	Irrigation water conservation benefits	Sedimentation benefits	Water quality benefits	Wildlife habitat benefits	Pheasant hunting benefits	Wetland creation benefits
	Average contract cost per acre[2]								
RE States	$207.26	$180.81	$359.52	$785.25	$311.97	$574.84	$511.91	$451.95	-
Non-RE States	58.94	27.77	54.88	142.03	73.97	100.26	39.8	50.09	-
	Standard deviation								
RE States	339.6	148.68	582.64	1,224.42	450.7	1,024.29	843.29	711.8	-
Non-RE States	174.27	46.13	90.11	235.21	143.27	225.64	54.77	121.33	-
	Minimum contract cost per acre[3]								
RE States	0.07	0.07	1.46	0.37	1.73	1.54	1.36	4.67	-
Non-RE States	0.1	0.2	0.7	0.14	0.1	0.3	0.51	0.51	-
	Maximum contract cost per acre								
RE States	1,513.08	507.88	2,154.53	5,354.42	1,759.24	4,968.55	3,407.07	1,958.09	-
Non-RE States	1,217.43	192.82	490.73	760.05	639.93	1,348.83	186.92	646.56	-

EQIP = Environmental Quality Incentives Program.

[1] Table includes costs associated with mitigating those physical effects for which monetary estimates of benefits exist.

[2] The costs per acre used to create these descriptive statistics were the contract cost per acre at the watershed level.

[3] Many contracts treated multiple resource concerns, and in those cases contract costs were divided among the resource concerns. This contributes to the small minimum cost per acre estimates.

Source: ERS analysis of NRCS contract data.

Table 4. Average benefits from conservation measures (in 2004 dollars).

	Benefits from reduced erosion from wind and water					Other benefits			
	Onsite soil productivity benefits	Air quality benefits	Sedimentation	Water quality	Grazing productivity	Irrigation water conservation	Wildlife viewing	Pheasant hunting	Wetland creation
	Dollars per acre								
RE States	$3.43	$1.82	$32.27	$16.83	$9.03	$2.61	$19.60	$0.06	$550.00
Non-RE States:									
Appalachia	3.83	-	24.14	14.43	12.00	-	23.47	0.16	50.00
Corn Belt	5.49	-	11.89	11.90	10.49	0.75	25.38	4.04	550.00
Delta	1.53	-	12.64	5.46	13.22	5.54	22.81	0.20	550.00
Mountain	0.09	3.87	$0.98	0.23	3.33	14.31	0.73	0.19	550.00
NE & Lake States	3.54	0.04	18.99	15.62	9.44	-	25.69	2.66	550.00
Northern Plains	0.74	1.68	2.64	1.87	5.24	2.98	3.55	1.92	550.00
Pacific	0.51	2.02	5.08	4.67	5.57	19.19	0.32	-	550.00
Southeast	1.79	-	19.78	7.54	12.64	2.86	23.31	-	550.00
Southern Plains	0.51	1.95	5.87	3.41	6.36	4.68	15.12	0.08	550.00

RE = Regional Equity Provision.

Note: Benefit measures are not available for Alaska, Hawaii, Puerto Rico, Pacific Basin Countries, and the Caribbean.

Source: ERS analysis of NRCS administrative data.

In WHIP, for the nine environmental categories considered, the Regional Equity provision generated losses of net benefits of about $1.8 million in 2005 and $4.9 million in 2006 (WHIP was not affected by the Regional Equity provision in 2004) (figure 10). The contracts enrolled in several States addressed different resource concerns between the 2 years, which contributed to this large difference. The measurable losses in net benefits in WHIP were also larger than those in EQIP in 2006, even though WHIP is a smaller program in terms of overall program funding. These cross-program effects might have been different, however, if data constraints had not prevented analysis of the impacts on net benefits for the full set of conservation activity in both programs. Benefit measures were available for more conservation activity in WHIP than in EQIP, so more EQIP activity was excluded from our analyses (see box, "Conservation Activities Linked to Environmental Benefits" and figure 6).

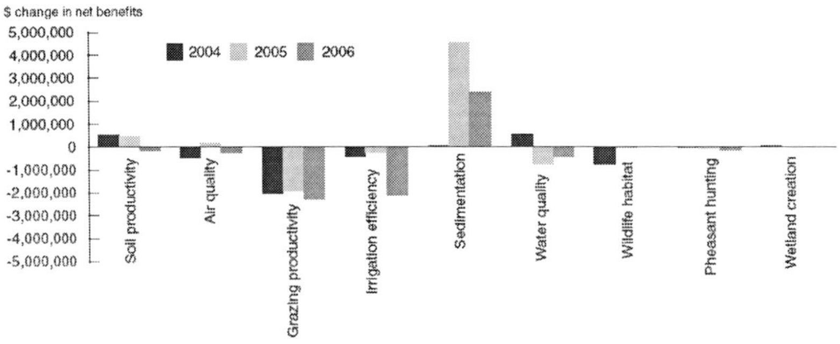

[1]Includes only the conservation activity funded through Environmental Quality Incentives Program (EQIP) in the continental U.S. for which estimates of the benefits of treatment are available in dollar terms.

Source: ERS analysis of NRCS administrative data.

Figure 9. Effects of the Regional Equity provision on selected net benefits in EQIP[1].

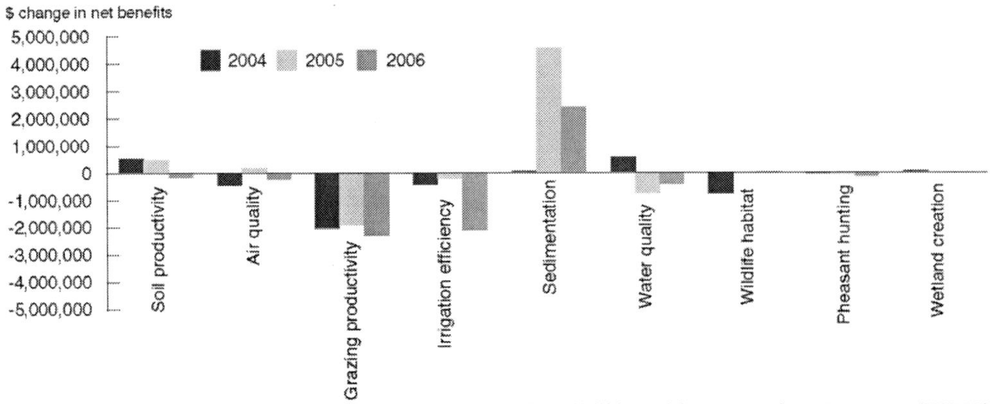

[1]Includes only the conservation activity funded through Wildlife Habitat Incentives Program (WHIP) in the continental U.S. for which estimates of the benefits of treatment are available in dollar terms.
Source: ERS analysis of NRCS administrative data.

Figure 10. Effects of the Regional Equity provision on selected net benefits in WHIP[1]

These analyses reveal that the Regional Equity provision resulted in a number of tradeoffs that were likely to affect the ability of the programs to provide environmental benefits cost effectively. They also reveal that the direction and magnitude of environmental and economic impacts of the Regional Equity provision are resource specific. For grazing productivity and water conservation issues in EQIP and wildlife and pheasant-hunting habitat in WHIP, the funding shifts across States reduced both the amount of land receiving treatment and the net benefits from treatment, at least during the 3-year period we studied. For most other resource issues, there were reductions in the amount of land receiving treatment, but the economic impacts were negligible or positive. The differing impacts across programs also suggest that the way an allocative policy is implemented may be an important determinant of the collective environmental and economic impacts.

The Regional Equity Provision Is Compatible with Distributional Goals of Working-Land Programs

The Regional Equity provision does not appear to work at cross-purposes with distributional goals in working-land programs. In addition to environmental, economic, and allocative goals, EQIP is designed to provide favorable treatment for certain types of operations and producers. Because targeted producers who participate in EQIP are not distributed equally among States and States do not always prioritize them for enrollment, the minimum funding allocations of the Regional Equity provision could have unintended impacts on participation by targeted producers. For example, by law, at least 60 percent of overall EQIP funding must be used for livestock-related conservation practices (USDA treats all practices implemented by livestock producers as livestock related (USDA, 2003)). Our analyses of payments by producer type suggest the 60-percent threshold would have been satisfied with or without the Regional Equity provision: Livestock operators received between 66 and 71 percent of EQIP funding annually after the Regional Equity funding reallocations occurred. Satisfying the Regional Equity provision reduced the number of livestock contracts, but only by an estimated 1.5 percent, or 500-700 contracts annually over the 2004-06 period.

Beginning farmers and farmers with limited financial resources also receive special attention in EQIP. The 2002 Farm Act authorized these producers to receive higher cost-share reimbursement rates (up to 90 percent of practice installation cost compared with a 75-percent maximum rate for other producers). The 2008 Farm Act authorized additional favorable payment terms, expanded eligibility for favorable terms to socially disadvantaged farmers, and required that beginning farmers and socially disadvantaged producers have access to at least 5 percent of program funds. Over the 2004-06 period, a larger proportion of contracts in the RE States were held by beginning and limited-resource producers, and while the Regional Equity reallocations allowed more of them to be enrolled, the impacts were small— an average of 100 additional contracts were written with these producers annually over the 2004-06 period. The funding shifts also had no appreciable impact on the overall EQIP funding received by beginning farmers, which averaged 11-12 percent of financial assistance annually over the 2004-06 period.

CONCLUSIONS

Allocative goals established by congressional legislation play an important role in the design of U.S. conservation programs for working lands. In conservation programs such as EQIP, WHIP, FRPP, and GRP, program funds are allocated to State program offices, where decisions are made about which resource problems will receive priority and which applications are accepted—either at the State or sub-State level. In EQIP, State offices can delegate these decisions, as well as the authority to choose payment rates and the conservation practices that will be funded, to local (sub-State) NRCS offices. This decentralized structure results in broad distribution of conservation funds, with producers in neighboring jurisdictions potentially having different incentives for enrollment.

How these legislative provisions are structured can have broad implications for which groups or regions are affected by program changes. In particular, policies that require *fixed* set-asides for particular regions result in disproportional impacts of program budget changes. For example, the Regional Equity provision guarantees that each State NRCS office is allocated funding equal to or greater than the fixed threshold ($12 million annually through EQIP, WHIP, FRPP, and GRP combined in the 2002 Farm Act, raised to $15 million in the 2008 Farm Act). If overall program funding were to fall, and there were no Regional Equity provision, each State NRCS office would experience a reduction in conservation program funding in proportion to the share of the budget that State normally receives. With a binding Regional Equity provision, however, a decline in overall program funding would be absorbed by the subset of States without protected funding. As program budgets rise, the impact of the Regional Equity provision on the States without protected funding is reduced. Alternatively, when set-asides are specified as a *proportion* of overall funding—such as the mandated 60-percent set-aside for livestock practices or the 5-percent set-aside for beginning farmers—the funding reserved for targeted groups is scaled up and down proportionally as budgets change.

When legislation applies to programs that already seek to achieve environmental, economic, distributional, and allocative goals, it is difficult to predict the direction and magnitude of potential tradeoffs when new legislation furthering an allocative goal is layered on top. Our analysis suggests the Regional Equity provision of the 2002 Farm Act resulted in a number of tradeoffs with other program goals. Meeting the requirement meant that each program agency had to shift funds among States. This changed the number and mix of producers who were accepted into the program in different States. While the provision did increase conservation funding for the dozen or so States that historically had received limited amounts through Federal conservation programs, it resulted in declines in the number of acres receiving EQIP funding to treat most environmental problems. The funding shifts also resulted in increased treatment costs per acre for many environmental projects funded by EQIP and WHIP. However, the shifts also resulted in more conservation activity being given to States where the benefits per acre from treatment were higher because more people benefited from the conservation activity undertaken by producers. In some cases, the increase in the per acre benefits was greater than the increase in treatment costs. For example, the Regional Equity provision decreased the net benefits (calculated as gross benefits minus costs) from improving grazing productivity by about $2 million annually over the 2004-06 period in EQIP, but increased the net benefits from reducing sedimentation problems by $4 million in 2005 and over $2 million in 2006.[16] While the Regional Equity provision's overall impact on benefits provided by the affected conservation programs could not be estimated because of data limitations, the patterns we found clearly show that the provision affects program benefits.

When program agencies have flexibility for implementing legislation, an understanding of the economic, environmental, allocative, and distributional impacts of alternative implementation options can guide the decision- makers' choice of a strategy that minimizes the losses of benefits from all affected programs. The Regional Equity provision reduced net benefits from addressing environmental problems in WHIP in the 2 years of our study when WHIP was subject to the provision's requirements, but the impacts in EQIP were more mixed. Impacts may not be correlated with program size; even though WHIP is smaller than EQIP in terms of overall funding, in 2006 the net losses to WHIP that we could measure with existing data were oneand-a-half times larger.

Policymakers face important decisions about how to use conservation program funds to deliver environmental benefits in a cost-effective manner. In conservation programs that are also designed to achieve distributional goals, it is not immediately apparent what impact a policy change to further allocative goals will have on a program's cost effectiveness or the amount of land receiving treatment. As analysis of the effects of the Regional Equity provision demonstrates, even when the amount of land receiving treatment decreases, net environmental benefits could increase if conservation activity shifts to places that provide more benefits. Because the environmental and economic characteristics of farms participating in conservation programs vary widely, understanding the characteristics of the contracts affected by a policy change—the conservation choices, treatment costs, and value of benefits provided by treatment—is necessary to determine the impacts of policy changes on all program goals in working-land programs.

APPENDIX A. ALLOCATIVE GOALS IN WORKING-LAND AND LAND PROTECTION PROGRAMS

In U.S. working-land and land protection programs, each program uses a unique set of design options that allow allocative as well as environmental, economic, and distributional goals to be met.

Table A.1. Indicators and weights used to allocate EQIP funds to States, 2006.

Indicators for resource issues	Weight
Acres of nonirrigated cropland	3.2
Acres of irrigated cropland	4.3
Acres of Federal grazing land	0.5
Acres of non-Federal grazing land	4.3
Acres of forestlands	1.1
Acres of specialty cropland	3.2
Acres of wetlands and at-risk species habitat	4.6
Acres of bodies of water	3.2
Livestock animal units	5.8
Animal waste generation	5.8
Waste management capital cost	3.5
Acres of American Indian tribal lands	3.3
Number of limited resource producers	5
Acres of grazing land lost to conversion	0.8
Air quality nonattainment areas	1.4
Acres of pastureland needing treatment	5.5
Acres of cropland eroding above T	6.2
Acres of fair and poor rangeland	6.2
Acres of forestlands eroding above T	1.4
Acres of cropland and pastureland soils affected by saline and/or sodic conditions	2.6
Miles of impaired rivers and streams	3.6
Potential for pesticide and nitrogen leaching	1.3
Potential for pesticide and nitrogen runoff	1.7
Ratio of livestock animal units to cropland	1.7
Number of concentrated animal feeding operations	2.8
Ratio of commercial fertilizers to cropland	0.9
Wind erosion above T	4.2
Phosphorus runoff potential	3.9
Riparian areas	0.8
Carbon sequestration	3.6
Coastal zone land	3.6
Total	100

Note: T is the soil loss tolerance factor (tolerable rate of soil erosion). EQIP = Environmental Quality Inventives Program.

Source: GAO, 2006.

The Environmental Quality Incentives Program (EQIP)

EQIP is USDA's major conservation program for working lands. It provides funding for conservation activities that address a wide range of resource issues, including soil quality, water quality, air quality, greenhouse gases, water and energy conservation, and wildlife habitat. All agricultural producers engaged in crop and livestock production are eligible to apply, as long as they comply with highly erodible land (Conservation Compliance) and wetland (Swampbuster) conservation requirements and do not exceed specified income levels.

EQIP has a long history of operating as a decentralized program with allocative goals. EQIP was enacted in 1996 and combined the conservation functions of four preexisting programs. One of its predecessors, the Agricultural Conservation Program (ACP), allocated program funds to State NRCS offices almost from its inception in 1936, with the goal of targeting funds to specific practices in all counties (Mercker, 1936). In 1996, the newly formed EQIP continued to allocate funds to State NRCS offices, based on conservation priorities. However, emphasis shifted to enrolling producers in locally determined "priority areas" facing the greatest environmental problems, rather than distributing enrollments across counties. In 2002, the practice of targeting enrollments to priority areas was eliminated, due primarily to producer dissatisfaction with the limited opportunities for enrollment outside priority areas (USDA, 2003).

In recent years, 31 indicators have been included in the mechanism used to allocate financial assistance across State NRCS offices. These indicators are of two types: (1) indicators measuring resource availability in the State (e.g., acres of grazing land and wetland) and (2) indicators measuring the extent of resource impairments (e.g., acres of highly erodible cropland, and the potential for pesticide runoff) (table A. 1) (U.S. GAO, 2006). NRCS weights each indicator based on an assessment of the relative importance of the corresponding resource issue. In 2006, the 31 resource indicators were each allocated between 0.5 and 6.2 percent of financial assistance funds. Funds for each indicator were then allocated to each State NRCS office based on the extent and impairment of the indicator in the State.

Since 2002, EQIP has given States wide flexibility in awarding contracts within their allocated funds, allowing States to target enrollments to meet their individual needs. Some States pool all applicants and make enrollment decisions at a State level, while others allocate all funds to sub-State NRCS offices that make the enrollment decisions. Some States adopt a mixture of these two approaches, retaining a portion of funds at the State level for specific resource concerns and allocating remaining funds to local NRCS offices. Each jurisdiction can have unique point systems for ranking and enrolling contracts, unique sets of conservation practices that are eligible for funding, and different payment rates (although since 2002, payment rates for conservation practice installation have been constrained by law to between 50 and 75 percent of practice cost, but up to 90 percent for beginning and limited-resource farmers). This means that producers in neighboring jurisdictions (even within the same State) can face different incentives for enrolling in EQIP. Since 2007, locally developed point systems have been required to assign a positive weight to national resource priorities (though the weights assigned to national issues need not be the same as those used to allocate funds to States). In addition to the relative weights of resource problems, many State and sub-State NRCS offices also consider contract cost in their rankings.

EQIP is subject to a number of Farm Act provisions for meeting distributional goals that target funds to specific producer types. At EQIP's inception, Congress mandated that 50 percent of funds be used to address resource issues on livestock and poultry operations, and this set-aside was raised to 60 percent in 2002. In 2002, the Farm Act granted favorable payment terms to beginning farmers and limited-resource producers, which were extended to socially disadvantaged farmers in 2008. The 2008 Farm Act requires that 5 percent of EQIP funds first be made available to beginning farmers, and another 5 percent to socially disadvantaged producers. States consider the set-aside requirements when designing their application processes, but because these requirements are easily met, there has been no national requirement for each State to set aside a particular percentage of funds.

Wildlife Habitat Incentive Program (WHIP)

WHIP was implemented in 1996 to assist landowners with developing or improving wildlife habitat on agricultural land. As in EQIP, NRCS allocates WHIP funds to State NRCS offices to achieve allocative goals and grants States flexibility to tailor the program to meet their individual needs. The funding allocated to each State is based on major wildlife-related resource priorities identified in Farm Act legislation (table A.2). Unlike EQIP applications, all WHIP applications are pooled and ranked at the State level, although, as with EQIP, States develop their own wildlife habitat priorities, sets of conservation practices that can receive funding, and application- ranking criteria. These ranking criteria include the expected longevity of the habitat to be created, parcel size and use, proximity to other, similar habitat (agglomeration benefit), cost, and any funding from other sources (USDANRCS, 2004b). As with EQIP, the State priorities and associated weights may differ from national priorities upon which the State allocations were based.

Grassland Reserve Program (GRP) and Farm and Ranch Lands Protection Program (FRPP)

GRP and FRPP provide payments to purchase easements to prevent the conversion of grassland (GRP) and farmland (FRPP) to other uses. Like EQIP and WHIP, these programs are decentralized and have made funds broadly available to producers in all States since program inception (in 1996 for FRPP, and 2002 for GRP). In both FRPP and GRP, State program offices use program funds and make enrollment decisions based on USDA State and national resource priorities. However, applications must be sponsored by a local entity willing to contribute a portion of the easement cost (i.e., a land protection program established by a State or local government or nongovernmental organization). These entities submit parcels for consideration that satisfy their own program goals, so ultimately the easements funded through FRPP and GRP reflect a combination of national, State, and local priorities.

In FRPP, NRCS has historically made program fund allocations to States based on the amount of farmland previously converted to developed uses in the State, the risk of further farmland conversions, estimates of land- owner interest in selling easements in FRPP, and the financial contributions and historical performance of cooperating entities that sponsor landowner enrollments.[17]

In GRP, the two agencies responsible for program administration, NRCS (for easements) and USDA's Farm Services Agency (for rental agreements), typically allocate program funding to States based on the number of grazing operations, pasture, and range acres under threat of conversion, and biodiversity considerations.[18] These funds are used to purchase easements, to fund long-term rental agreements to maintain grassland cover, and to provide cost- sharing assistance to restore or install grassland.

Table A.2. Factors and weights used to allocate WHIP funds to States.

Resource concerns	Weight
At-risk wildlife species	20
At-risk wildlife species native habitat	20
Non-Federal acres	10
Total non-Federal acres enrolled in CRP, GRP, WRP	10
Declining grasslands needing restoration	8
Declining forestland needing restoration	7
Declining wetlands needing restoration	7
Aquatic species	8
At-risk aquatic species	10
Total	100

WHIP = Wildlife Habitat Incentives Program; CRP = Conservation Reserve Program; GRP = Grassland Reserve Program; WRP = Wetlands Reserve Program.

Source: GAO, 2006

Appendix B. Marginal Contracts

It may be tempting to estimate the impacts of policy changes based on aggregate measures like the "average" contract. However, as the following analyses reveal, the characteristics of conservation program contracts vary by priority. The underlying differences suggest that estimating impacts of policy changes based on the resources or size of the area treated by the "average" contract can be misleading. We use a small (10-percent) increase in conservation program budgets as an example to demonstrate how different estimated impacts might be.

As program budgets or other program factors change, funds allocated to each State will change. Marginal contracts (the lowest priority contracts that received program funding) and marginally rejected applications (unfunded applications with scores just below those of the marginal contracts) will be most affected by these budget changes. Figure B. 1 illustrates how hypothetical changes in State allocations affect marginal contracts and applications that offer to treat two problems: erosion and grazing productivity. Each dot represents a score based on the application's expected reduction in erosion and improvement in grazing productivity, and each dot's coordinates in the figure depend on the extent to which the producer agrees to treat each problem. The solid line represents the amount of erosion reduction and grazing improvements that can be funded with the available budget; it can be thought of as a cutoff line. For simplicity, unit costs of treating erosion and grazing lands are assumed to be the same for all producers. The olive-green dots have the highest scores and are funded contracts (accepted offers), given the available budget. The black dots have the lowest scores and are rejected applications. The gray dots, those closest to the cutoff line, are the marginal contracts and marginally rejected applications. If the budget increases slightly (cutoff line moves closer to the origin, as in State A's figure), applications with the next highest scores would be funded—they become the marginal contracts. When budgets decline slightly (as in State B's figure), the marginal contracts no longer receive funding—they become "marginally rejected." The positions of the dots vary between the two States depicted in the figure, reflecting the idea that the mix of problems that producers agree to treat varies across States. The locations and slopes of the budget lines reflect, respectively, the variation in the size of State budgets and the different priorities States place on treating different problems. The impacts that relatively small program changes will have on program outcomes will depend largely on the characteristics of marginal contracts and marginally rejected applications in each State—the number of acres producers agree to treat, the payment producers are willing to accept, and the environmental benefits they are willing to provide. However, when legislation represents a major overhaul of a program, the incentives that producers face will change, and consideration of how the set of producers who apply to a program may change in response (the number and positions of the dots in the graphs) will be important.

When conservation programs are decentralized and allow for multiple jurisdictions to make enrollment decisions, as in EQIP and WHIP, multiple ranking tools will be at work in each program. In EQIP, many ranking tools can exist within each State. This decisionmaking structure means that marginal contracts, and marginally rejected applications, will exist for each application pool—potentially yielding a set of marginal contracts and applications that vary widely in terms of resource problems and costs of treatment.

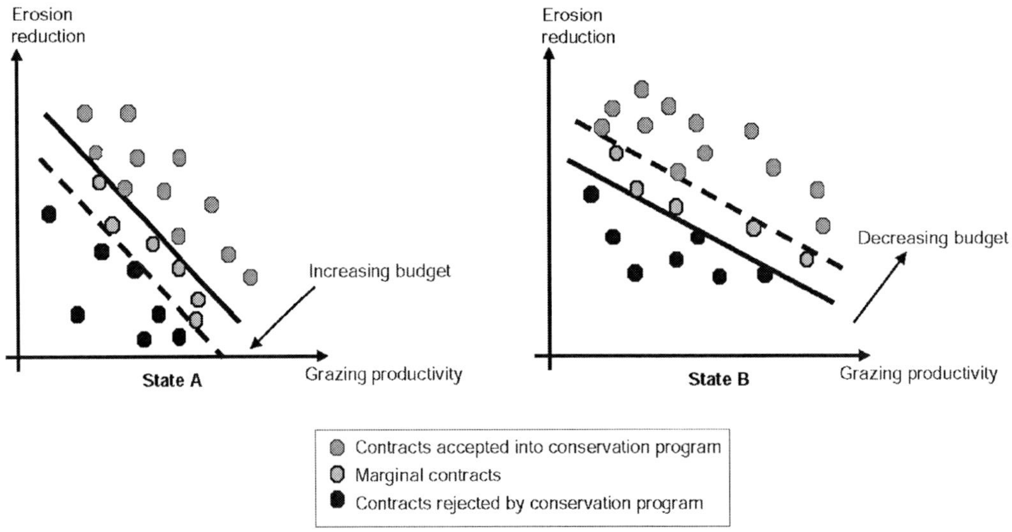

Source: USDA, Economic Research Service. Adapted from Cattaneo et al., 2006.

Figure B.1. Increasing program budgets allow more contracts to be funded.

How Do Marginal Contracts Differ From Higher Priority Contracts?

Because of the diversity in the resource problems that producers face, the flexibility producers have in deciding which problems to treat, and the few limitations on eligibility for working-land conservation programs, contracts enrolled in these programs can vary in many ways. State agencies administering conservation programs also affect the characteristics of enrolled contracts through the conservation priorities the agencies establish. The process for prioritizing applications in a conservation program suggests that differences are likely to exist between those of higher priority for enrollment and those of lower priority. An obvious difference is that higher priority contracts are more likely to address higher priority resource concerns. They may also be more likely to address these issues at lower cost, at least in jurisdictions that consider per acre treatment costs in the prioritization process. Whether the contract addresses concerns arising from livestock operations, or aids particularly disadvantaged producer types, may also differ by contract priority.

Comparing the characteristics of marginal to higher priority contracts in EQIP and WHIP—the financial and cost-share assistance received, the types of practices implemented, and the resource concerns addressed—illustrates how these contracts vary in programs treating environmental problems on working lands. We identified contracts as marginal or not using the contract's ranking score and information about which contracts were pooled and ranked against each other in the selection process, information which is included in the program administrative data maintained by NRCS.[19] We defined low-priority contracts as those that ranked the lowest in each funding pool, which, in total, comprised 10 percent of program funds. Highest priority contracts were those that ranked the highest and comprised 10 percent of program funds.

Marginal contracts differ from higher priority contracts in a number of ways. For example, marginal contracts in EQIP received significantly smaller (in a statistical sense) contract payments compared with the highest priority contracts (table B.1). A regional-level analysis reveals that this pattern held, almost without exception, across USDA production regions and over the 3-year period 2004-06. Contract payments on marginal contracts were lower because these contracts tended to receive both lower rates of cost-share assistance and smaller financial assistance payments. In EQIP, participants receive cost-share assistance to help offset part of the cost of installing structural or vegetative practices, and financial assistance to adopt more environmentally friendly management practices.[20] In 2006, the average rate for cost-shared practices was 57 percent and 61 percent for marginal and high-priority contracts, respectively (table B.1).

Table B.1. Characteristics of EQIP contracts by priority, 2006.

Characteristic	Marginal contracts	Highest priority contracts	Difference
Average contract payment, dollars	$11,795	$15,720	-$3,925**
Average percentage rate for cost-shared practices	0.57	0.61	-0.04**
Average incentive payment, dollars	$2,707	$3,231	-$524**
Number of practices, average	2.63	3.4	-0.77**
Size of commonly adopted practices:			
Fencing, feet	4,216	5,992	-1,776**
Nutrient management, acres	400	341	59**
Pasture and hay planting, acres	51	79	-28**

**Difference is significant at a 95% confidence level.

EQUIP = Environmental Quality Incentives Program.

Source: USDA-NRCS administrative data for EQIP, 2006.

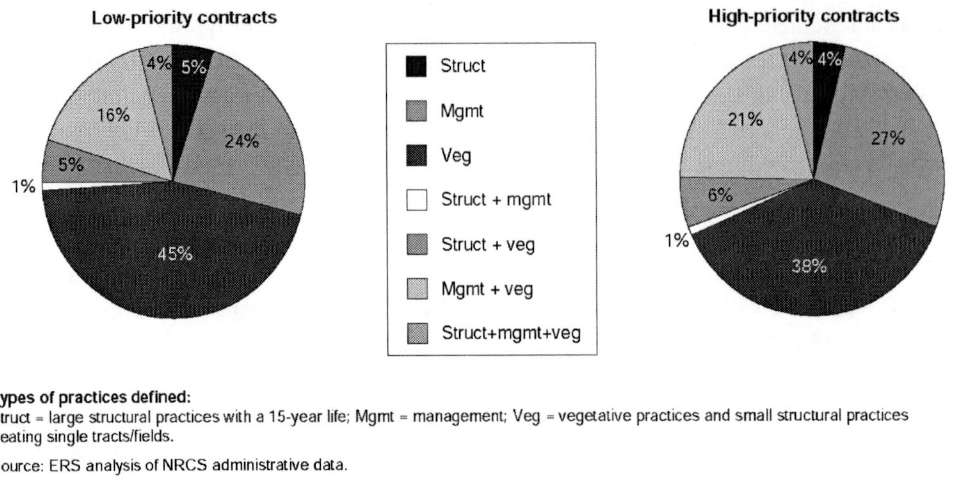

Source: ERS analysis of NRCS administrative data.

Figure B.2. Percent of EQIP contracts by practice type and priority.

Low- and higher priority contracts also differ based on the amount of land that receives treatment and types of practices producers agree to implement. Analysis of 2006 EQIP contract data reveals that the typical lowest priority contract treated fewer acres compared with higher priority contracts, ranging from 5 percent fewer acres treated for water quality issues to 84 percent fewer acres for fish and wildlife issues. Most EQIP contracts involve implementation of a single *type* of practice—in 2006, about 70 percent of all EQIP contracts implemented only structural, vegetative, or management practices (figure B.2). Of these contracts, low-priority contracts tended to implement strictly structural or vegetative cost-shared practices more frequently than higher priority contracts did (50 percent versus 41 percent), while more of the highest priority contracts included combinations of practice types. Installing conservation "systems" involving multiple practices may be the most efficient way to treat a variety of resource problems (Lerch et al., 2005; Berry et al., 2003).

In recent years, policymakers have sought to improve participation rates by certain producer groups in EQIP. For example, the 2002 and 2008 farm legislation provided for higher levels of financial assistance and other measures to reduce barriers to participation by beginning farmers and limited-resource producers (the 2008 Farm Act includes favorable terms for socially disadvantaged farmers, as well). The 2006 enrollment data suggest beginning and limited-resource farmers and ranchers tend to get enrollment priority in EQIP. When viewed from a national level, 68 percent of payments on high-priority contracts are made on contracts held by beginning farmers and ranchers, while just 10 percent of payments to low priority contracts are received by these farmers (figure B.3). Payment patterns are similar for limited- resource producers.

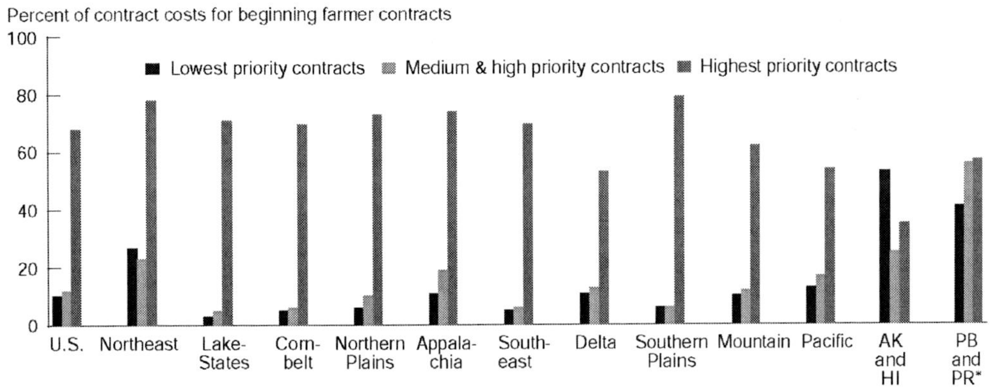

PB = Pacific Basin Countries; PR = Puerto Rico.
EQIP = Environmental Quality Incentives Program.
Source: ERS analysis of NRCS administrative data.

Figure B.3. Proportion of contract payments to beginning farmers in EQIP, 2006

Marginal Contracts: Implications of Differences for Program Analysis

Knowledge about the many ways in which lower and higher priority contracts differ can be particularly important for understanding the environmental, economic, distributional, and allocative impacts of policy changes, as well as for estimating how program administration requirements (such as staffing and provider needs for technical assistance) might change in response. Taking this variation into account is important even for relatively small program changes. For example, in EQIP, if estimates of impacts of a 10-percent budget increase on contract numbers were based on the "average" contract size rather than the smaller size of marginal contracts, the number of additional contracts enrolled would have been understated by 42 percent. Further, estimates of the number of acres receiving treatment for plant conditions, soil erosion, soil conditions, and fish and wildlife problems would be overstated by more than 40 percent, because marginal contracts tend to treat fewer acres than the average contract.

Understanding differences among contracts by producer type is also important, especially when certain producers require additional technical or other types of Government assistance. The average EQIP payment to all livestock producers was about $21,120 in 2006, while payments for livestock producers with marginal contracts were about half as large. If the impacts of a 10-percent budget increase were based on the size of the "average" contract and livestock contracts continued to receive 66 percent of any additional funding (the proportion of all funding received by this producer group nationally), the number of new livestock contracts that could be enrolled with a 10-percent budget increase would be underestimated by about 45 percent. The number of contracts with beginning and limited resource farmers would also be underestimated by 25 percent, if similar assumptions about contract sizes were made.

APPENDIX C. METHODOLOGY FOR ESTIMATING ACRES RECEIVING TREATMENT AND THE COSTS PER ACRE OF TREATING RESOURCE CONCERNS

The EQIP and WHIP data we used to estimate treatment costs included the cost of implementing each practice, identifiers for the fields where each practice was to be installed, indicators of physical effects to be treated, and information on how many acres were receiving treatment under the contract, as identified by local NRCS field offices. We used these locally determined links between costs and environmental problems to estimate watershed-level costs of treatment. The advantage of using the locally determined links is that they incorporate the natural variation in practice costs across different States, which depend on local priorities and circumstances.

Agricultural producers address resource concerns by implementing conservation practices that mitigate the underlying physical effects of production. For example, pasture and hay planting and fencing are practices often used to mitigate overgrazing, a physical effect that addresses the resource concern of the productivity of grazing land. Often more than one practice is used to address a physical effect. Also, some practices, like fencing, can mitigate multiple physical effects (e.g., overgrazing and streambank erosion) that address more than one resource concern (such as grazing productivity and water quality). In developing our cost estimates, we accounted for both the multiple-practice and multiple-resource concerns evident in the data.

The study's primary goal in developing the cost estimates was to do so in such a way that the costs could ultimately be compared to benefits for addressing particular resource concerns. Three factors were important: (1) linking the physical effects to resource concerns (such as water quality or grazing productivity) to develop cost estimates of addressing the concern; (2) identifying which costs were incurred to mitigate which physical effects (such as soil erosion or grazing land health); and (3) grouping practices based on the years of benefits that they were likely to generate.

In fiscal years 2004 and 2005, NRCS field offices could assign practices to as many as 76 physical effects, depending on the environmental problem the practice was expected to treat. In 2006, NRCS changed and expanded the list to include 82 physical effects. For our analysis, we grouped the physical effects by resource-concern categories for which we had benefit estimates (see Appendix D for a description of benefit estimates). For 2004-2005, 28 of the 76 physical effects were mapped to resource concern categories for which benefit estimates existed; in 2006, 17 of 82 physical effects were mapped. Table C. 1 identifies which subsets of physical effects linked to which resource concerns. Although the physical effects receiving treatment differed somewhat during the study period, we reviewed the data and found reasonable consistency across years between the practices that were associated with particular resource concerns.[21]

In some cases, the practice-to-physical-effect links in the data appeared tenuous—for example, ponds were identified as reducing wind erosion in some contracts. In such cases, we imposed restrictions on the data. We paid particular attention to large structural practices (e.g., creating ponds, waste storage facilities, etc.) that were associated with unlikely physical effects and that would have the most significant effects on net benefit estimates.

Table C.1. Physical effects included in benefit categories.

Resource concerns	Physical effects receiving treatment	
	Fiscal 2004, 2005	Fiscal 2006
Benefits from reduced erosion from wind and water		
Onsite soil productivity benefits	Sheet and rill erosion reduction, reduction in irrigation-induced soil erosion	Sheet and rill erosion reduction, reductionin irrigation-induced soil erosion
Air quality benefits	Reduction in wind erosion and particulate pollution	Reduction in wind erosion and particulate matter less than 10 micrometers in diameter
Sedimentation reduction benefits	Reduction in damage from sediment deposition and erosion from roadsides and construction sites; reduction in	Reduction in damage from sediment deposition and erosion from road sides and construction sites; reduction in shoreline
	shoreline and streambank erosion; reduction in sediment deposition	and streambank erosion; improvement in capacity of conveyances from reducing sediment deposition
Water quality benefits	Reduction in surface water salinity, sediment and eutrophication, heavy metals, pesticides, acid rain problems, and temperature extremes; reduction in bioaccumulation of toxins; improvement in water for fish and wildlife	Reduction in surface water salinity, sediment and turbidity, heavy metals, pesticides, petroleum and harmful temperatures; rectifying inadequate water for fish and wildlife
Grazing productivity benefits	Reduction in noxious and invasive plants, improvement in rangeland site stability	Improvement in quantities and quality of feed/forage/plants, reduction in noxious and invasive plants, improvement in rangeland site stability and hydrologic cycle
Irrigation efficiency benefits	Improvement in water supply for irrigated land	Improvement in efficiency of water use onirrigated land
Wildlife viewing benefits	Improving population imbalances and species diversity, reduction in loss/degradation of riparian habitat/vegetation, reduction of nonindigenous plant or animal species	Improving population imbalances, food, space, and populations of threatened and endangered species and species of concern
Hunting benefits	Reduction in loss/degradation of forest or grass cover and habitat fragmentation	Improving cover/shelter for wildlife and reducing habitat fragmentation
Wetland creation benefits	Reduction of significant hydrological modification	Reduction in excessive subsurface water

Source: USDA-NRCS administrative data for Environmental Quality Incentives Program (EQIP) and Wildlife Habitat Incentives Program (WHIP).

In addition to restrictions on links between large structural practices and particular physical effects, the primary restrictions we imposed were on the types of practices related to wetland creation and grazing productivity issues (e.g., we assumed practices funded through WHIP were not intended to mitigate grazing land productivity issues).

To identify the costs of mitigating physical effects, we used the contract data to link practices both to the fields on which they were installed and to the indicators of physical effects expected to be mitigated by those practices (we assumed that practices completely addressed the identified physical effects on all fields where the practice was installed). This yielded cost estimates that varied by practice, year, and field level. Most practices were installed on a single field, but in some cases they were installed on many fields. In 2004 and 2005, 79 percent and 87 percent of EQIP contract items were coded to three or fewer physical effects, with 46 percent and 66 percent of contract items coded to just one physical effect in those years.

In generating contract-specific cost estimates, we identified whether the physical effect was mitigated using one of seven different practice types or combinations of types: large structural practices, management practices, vegetative practices, large structural plus management practices, large structural plus vegetative practices, management plus vegetative practices, or all three types (large structural plus vegetative plus management practices).[22] Distinguishing by practice type allowed us to match costs to varying streams of benefits, based on the varying lengths of the useful lives of the practices.

When estimating the per acre cost of addressing a resource concern, we had to account for the fact that multiple practices could be applied to the same field without double-counting the acres treated. In cases where practices addressed a single resource concern, we estimated the contract cost per acre (CPA) of addressing resource concern *r* with all practices of type *P* by summing up the costs of each practice c^P over the *t* years implemented, and dividing by the number of acres *a* the practices are estimated to treat on *j* fields (contract subscripts are suppressed to simplify notation):

$$CPA^{rP} = \frac{\sum_{t}\sum_{p=1}^{P} c_t^{rp}}{\sum_{j^P} a_{j^P}}$$

Ideally, data would be available on the number of acres treated by individual practices (a). Because acreage data are only available at the contract level, we estimated the acreage treated by each practice on field j (a_j) by allocating the total acres treated by the contract across all the fields that were subject to treatment by any practice. Some practices within a contract may be implemented on fewer fields than others, so total treated acres (*TTA*) are divided by the maximum number of fields receiving treatment by *any* practice in the contract:

$$a_j = TTA/argmax(j),$$

which assumes all fields treated by a contract are of equal size.

The above formulas assume that practices are implemented to address a single resource concern. In some cases, practices address multiple physical effects that map to multiple resource concerns. These cases raise the question of how to allocate practice costs across multiple concerns. Most practices are implemented to address a primary physical effect, so a reasonable strategy would be to allocate some costs to that effect. For this analysis, we allocated practice costs by apportioning them equally across the physical effects being mitigated by the practice.

As a sensitivity check, we also analyzed how outcomes would change if practice costs and treated acres were allocated entirely to the "primary" physical effect treated by the practice. We estimated the primary effect based on which physical effect each practice treated most frequently. We calculated the primary physical effect by practice within each State, which maintained spatial variation across States in the practice-physical effect mappings. With costs for addressing physical effects calculated this way, our estimates of the impacts of the Regional Equity provision on net benefits differed from our results when we apportioned costs equally across all physical effects being addressed by the practice, but not in any predictable way. The net benefits for some resource concerns increased, while others decreased up to several hundred thousand dollars. This suggests that assumptions about how costs are allocated when multiple resource concerns are treated can affect the outcomes.

Accounting for Outliers in the Data

One issue that arose in calculating the cost per acre for each resource concern was that for some contracts, the cost per acre estimates were unreasonably high or low. In large part, these outliers were due to inconsistencies in how field offices entered data on the acres that received treatment. The primary problem was that the number of acres on which the practice was installed was reported as "treated acres." For example, a riparian buffer *installed* on 0.2 acres was listed as *treating* 0.2 acres, when it actually treated (prevented soil erosion from leaving) a 20-acre field. This type of inconsistency resulted in cost per acre estimates that were unreasonably high. In other cases, certain practices were recorded as treating tens of thousands of acres. While possible, practices treating such large acreages are not typical, and they result in unreasonably low cost per acre estimates. Because calculations of average per acre costs at aggregated levels are affected by outliers, we used median per acre costs, or the per acre estimates for which half the observations lie below the estimate and half lie above (when the per acre cost estimates are arranged in order of magnitude).

Summarizing per Acre Costs to the Watershed Level

To be consistent with benefit estimates that varied across watersheds (8-digit Hydrologic Unit Codes, or HUCs delineated by the U.S. Geological Survey), we summarized the cost data to the watershed level. For each resource concern *r* and practice group *p*, we calculated the watershed level cost per acre as the median cost per acre of contracts in that HUC that treat resource concern *r* with practice group *p*. We also used the median number of acres treating resource concern *r*, multiplied by the number of contracts treating resource concern *r*, to estimate the number of acres receiving treatment.

APPENDIX D. BENEFIT ESTIMATES

The benefits from addressing resource concerns were estimated for each contract. We estimated these by linking the physical effects addressed in each contract to different benefit categories for which benefit values are avail-able from the literature (see Appendix C, table C. 1, for a description of how physical effects were mapped to benefit categories). These categories include benefits associated with erosion reduction (soil productivity, dust, sedimentation, water quality), improved water management (irrigation efficiency), grazing land productivity, and habitat creation and improvement (hunting, wildlife viewing, wetlands). Most of the procedures and data behind the benefit estimates are the same as those used in the benefit assessment of the Conservation Security Program (USDA, 2004a).

Benefit estimates for each category were available at several different geographic levels (Hydraulic Unit Code, State, farm production region, or national). GIS procedures were used to aggregate benefits to the State level. Benefits were placed on a per ton of erosion or per acre basis, depending on the category. Benefits from Environmental Quality Incentives Program contracts were estimated by applying the per unit values to the amount of erosion reduced or the number of acres treated.

Benefits from Reduced Erosion

Benefits from reduced water and wind erosion were obtained from Hansen and Ribaudo (2008), who provide economic measures of soil conservation benefits for a variety of benefit categories, representing benefits from reduced sedimentation, improved water quality, improved air quality, and improved soil productivity. Benefits are provided in terms of dollars per ton of erosion reduced, by county and the U.S. Geological Survey's 8-digit Hydrologic Regions (although the benefits for sub-State units are often assigned the per ton benefit estimated at a State or regional level).

In the case of erosion reductions, contract data did not contain any information on baseline erosion rates or the expected change in the rates. We assumed that practices adopted for the purpose of reducing erosion would achieve a reduction equal to the difference between the HUC-wide average erosion rate (tons per acre) on cropland eroding at T and more than T. T is the "acceptable" level of erosion as defined by NRCS (in tons per acre) and a standard target for erosion control practices. Cropland already at or near T would not need conservation practices to control erosion. We estimated HUC-wide average erosion rates with data from the National Resources Inventory on cropland with an erosion rate greater than T, and used this information to convert the benefits from reducing erosion on a per ton basis to a per acre basis.

Benefits from Improved Water Management

Benefits from improved water management represent the reduced cost to the farmer of water used in irrigation. Estimates of annual per acre water savings were obtained from NRCS (USDA-NRC S, 2004a). In that study, the assumption was made that improved water management reduces water use by about 13 percent. Potential per acre benefits were calculated using data on irrigation water use, water source (surface or ground), and irrigation water costs.

Benefits from Improved Grazing Productivity

Improved land management can increase the productivity of grazing lands. We used per acre productivity benefit estimates developed by NRCS at the multistate level for an analysis of the Conservation Security Program (USDA- NRCS, 2004a). Benefit estimates were developed by varying the national per acre values ($15.01 for pasture, $5.00 for grazing) across HUCs, based on relative soil productivity measures obtained from the NRI. Separate benefits were calculated for rangeland and pastureland. Where watersheds contained both rangeland and pastureland, we calculated the benefits for the HUCs as the weighted average benefit. We used the number of acres in rangeland and pastureland in the watershed, estimated using 1997 NRI data, as weights.

Benefits from Improved Hunting Opportunities

Improving wildlife habitat can produce hunting and wildlife-viewing benefits. We obtained estimates of the value of improved pheasant hunting from a study by Feather, Hellerstein, and Hansen (1999). Their study derived estimates by applying a Random Utility Model of pheasant hunter behavior (Hansen, Feather, and Shank, 1999). Recreation data for the model were from the 1991 National Survey of Fishing, Hunting, and Wildlife-Associated Recreation, and land-use data were taken from the 1992 NRI. For purposes of the present study, we reduced the benefit estimates by 50 percent, on the assumption that land remaining in production would generate half the benefits of land that is retired in the Conservation Reserve Program (CRP) (USDA-NRCS, 2004a).

Benefits from Wildlife Viewing

The 1999 study by Feather, Hellerstein, and Hansen estimated the impact of the CRP on nonconsumptive, wildlife-oriented recreation. In that study, data from the Hunting and Fishing Survey (U.S. Department of the Interior and U.S. Department of Commerce, 1993, 1997) were used to estimate how improvements in habitat affected the number of trips people took to view wildlife and the value of a trip in each region. Results were reported on a basis of per acre of-land in the CRP. Several studies have shown that a number of practices funded through EQIP have benefited wildlife, including irrigation water savings practices (which benefited the arctic grayling fish species in Montana and Wyoming), conservation tillage (which benefited nesting birds) and rangeland practices (which benefited the prairie grouse) (e.g., Riley 2004; Martin and Forsyth 2003; Lokemoen and Beiser, 1997; and Jahn and Schenck 1991).

Benefits from Wetland Creation

Wetland restoration can provide habitat for wildlife species that support recreational activities, such as hunting and fishing, and commercial activities such as fishing and trapping. Wetlands also have a nonuse value, primarily for the viewing of wildlife that depend on wetlands for nesting, cover, and food. We applied annualized recreation, commercial, and nonuse values for wetlands of $550 per acre based on values reported in the literature (Heimlich et al., 1998).

Applying Benefit Estimates to Practices

Table 4 reports the annual per acre benefit estimates used to estimate the impacts of the Regional Equity provision on net benefits. We took into account that the practices implemented under an EQIP or WHIP contract produce benefits over multiple years, and using a 7-percent discount rate, we calculated the present value of the benefit stream for each contract. We allowed the length of the benefit stream to depend on the types of practices (management or structural) and the length of the contract (Table D. 1). We applied estimates of practice lives reported by NRCS (2004): structural practices generate a 15-year stream of benefits, vegetative practices (buffers) generate a 10-year stream, and management practices generate a 4-year stream. No benefits were assumed to be generated in the initial year of implementing the practice establishment. An implicit assumption with this approach is that installed practices mitigate physical effects to a nondegradation level. If they do not, the benefits ascribed to the practices will be overstated.

Many participants agree to implement a suite of practices to mitigate a single physical effect. We assumed that all the practices would need to be in place in order for the contract to generate all of the benefits from conservation treatment. If, for example, both structural and management practices were used to mitigate a physical effect on the same land unit, we assumed that both types of practices were necessary to treat the land to a nondegradation level. For contracts that combined structural and management practices, we assumed that benefits from structural practices were reduced by 25 percent once the management practices ended after 3 years. For contracts that combined structural and vegetative practices, we assumed that benefits from structural practices were reduced by 25 percent once the vegetative practices finished their useful life after year 10. If both vegetative and management practices were implemented, we assumed that the benefits from vegetative practices were reduced by 25 percent once the management practices were no longer implemented (after 4 years). If all three types of practices were implemented, we assumed that total benefits were reduced by 15 percent between years 4 and 10 when the management practice was no longer implemented, and by 25 percent when only the structural practice was in place after year 10.

Table D.1. Portion of benefits generated by contract type, by year.

	Practice type						
	Struct.	**Mgmt.**	**Veg.**	**Struct. + Mgmt.**	**Struct. + Veg.**	**Mgmt.+ Veg.**	**Struct+Mgmt +Veg**
Benefits in year 1	0	0	0	0	0	0	0
Benefits in year 2	1	1	1	1	1	1	1
Benefits in year 3	1	1	1	1	1	1	1
Benefits in year 4	1	1	1	1	1	1	1
Benefits in year 5	1		1	0.75	1	0.75	0.85
Benefits in year 6	1		1	0.75	1	0.75	0.85
Benefits in year 7	1		1	0.75	1	0.75	0.85
Benefits in year 8	1		1	0.75	1	0.75	0.85
Benefits in year 9	1		1	0.75	1	0.75	0.85
Benefits in year 10	1		1	0.75	1	0.75	0.75
Benefits in year 11	1			0.75	0.75		0.75
Benefits in year 12	1			0.75	0.75		0.75
Benefits in year 13	1			0.75	0.75		0.75
Benefits in year 14	1			0.75	0.75		0.75
Benefits in year 15	1			0.75	0.75		0.75

Note: A value equal to 1 means full benefits are assumed to be generated in that year.
Types of practices defined:
Struct = Large structural practices with a 15-year life.
Mgmt = Management.
Veg = Vegetative practices and small structural practices treating single tracts/fields.
Source: ERS analysis of NRCS contract data.

REFERENCES

American Farmland Trust. (2008). "*Federal Farm and Ranch Lands Protection Program Final Allocations.*" FIC Fact Sheet, November. (Background survey; not cited in text.) (http://www.farmlandinfo.org/docu-ments/3 11 59/2008_FRPP_Final_Allocations.pdf).

Berry, J. K., Delgado, J. A., Khosla, R. & Pierce, F. J. (2003). "Precision conservation for environmental sustainability." *Journal of Soil and Water Conservation, 58(5)*, 332-39.

Bockstael, N. E., McConnell, K. E. & Strand, I. E. (1989). "Measuring the Benefits of Improvements in Water Quality: The Chesapeake Bay," *Marine Resource Economics, 6(1)*, 1-18.

Caswell, M., Fuglie, K., Ingram, C., Jans, S. & Kascak, C. (2001). *Adoption of Agricultural Production Practices: Lessons Learned from the U.S. Department of Agriculture Area*

Studies Project. Economic Research Service, U.S. Department of Agriculture, AER-792, January, 116 pp. (www.ers.usda.gov/publications/aer792/).

Cattaneo, A., Claassen, R., Johansson, R. & Weinberg, M. (2005). *Flexible Conservation Measures on Working Land: What Challenges Lie Ahead?* Economic Research Service, U.S. Department of Agriculture, ERR-5, June, 48 pp. (www.ers.usda.gov/publ ications/ERR5).

Cattaneo, A., Hellerstein, D., Nickerson, C. & Myers, C. (2006). *Balancing the Multiple Objectives of Conservation Programs*. Economic Research Service, U.S. Department of Agriculture, ERR-19, May, 58 pp. (www.ers. usda.gov/publications/ERR19/).

Claassen, R., Hansen, L., Peters, M., Breneman, V., Weinberg, M., Cattaneo, A., Feather, P., Gadsby, D., Hellerstein, D., Hopkins, J., Johnston, P., Morehart, M. & Smith, M. (2001). *Agri-Environmental Policy at the Crossroads: Guideposts on a Changing Landscape*. Economic Research Service, U.S. Department of Agriculture, AER-794, January, 72. (www.ers.usda.gov/publications/aer794/).

Feather, P., Hellerstein, D. & Hansen, L. (1999). *Economic Valuation of Environmental Benefits and the Targeting of Conservation Programs: The Case of the CRP*. Economic Research Service, U.S. Department of Agriculture, AER-778, May, 64.

Hand, M. & Nickerson, C. (2008). "Conservation Program Provision May Have Limited Impact on Underserved Farmer Participation," *Amber Waves*, *Vol. 6*, Issue 5, November, 7.

Hansen, L. & Ribaudo, M. (2008). "*Economic Measures of Soil Conservation Benefits: Regional Values for Policy Assessment*." Economic Research Service, U.S. Department of Agriculture, TB-1922, September, 31, (www.ers.usda.gov/publications/TB1922/).

Hansen, L., Feather, P. & Shank, D. (1999). "Valuation of Agriculture's Multi-site Environmental Impacts: An Application to Pheasant Hunting." *Agricultural and Resource Economics Review, 28(2)*, 199-207, October.

Heimlich, R. E., Weibe, K. D., Claasen, R. & House, R. M. (1998). W*etland and Agriculture: Private Interests and Public Benefits*. Economic Research Service, U.S. Department of Agricultlure, AER-765, 94.

Hopkins, J. & Johansson, R. (2004). "Beyond Environmental Compliance: Stewardship as Good Business," *Amber Waves*, *Vol. 2*, Issue 2, April, 30-37. (www.ers.usda.go v/Amberwaves/April04/Features/ BeyondEnvironmental.htm).

Jahn, L. R. & Schenk, E. W. (1991). "What sustainable agriculture means for fish and wildlife," *Journal of Soil and Water Conservation, 46(4)*, 251-54.

Key, Nigel, D. & Jonathan, D. Kaplan. (2007). "Multiple Environmental Externalities and Manure Management Policy," *Journal of Agricultural and Resource Economics, 32(1)*, 115-34, April.

Lambert, D., Sullivan, P., Claassen, R. & Foreman, L. (2006). *Conservation- Compatible Practices and Programs: Who Participates?* Economic Research Service, U.S. Department of Agriculture, ERR- 14, February, 43, (www.ers.usda.gov/publica tions/ERR14/).

Leggett, C. G. & Bockstael, N. E. (2000). "Evidence of the Effects of Water Quality on Residential Land Prices," *Journal of Environmental Economics and Management, 39(2)*, 121-44, March.

Lerch, R. N., Kitchen, N. R., Kremer, R. J., Donald, W. W., Alberts, E. E., Sadler, E. J., Sudduth, K. A., Myers, D. B. & Ghidey, F. (2005). "Development of a conservation-oriented precision agriculture system: Water and soil quality assessment," *Journal of Soil and Water Conservation, 60(6)*, 411- 20, November-December.

Lokemoen, J. T. & Beiser, J. A. (1997). "Bird use and nesting in conventional, minimum-tillage, and organic cropland," *Journal of Wildlife. Management, 61(3)*, 644-55.

Martin, P. A. & Forsyth. D. J. (2003). "Occurrence and productivity of songbirds in prairie farmland under conventional versus minimum tillage regimes," *Agriculture, Ecosystems and Environment, 96(1-32)*, 107-17.

Mercker, A. E. (1936). "The Agricultural Conservation Program," *American Journal of Potato Research, 13(5)*, 113-17, May.

Nickerson, C. & Hand, M. (2009). Participation in Conservation Programs by Targeted Farmers: Beginning, Limited-Resource, and Socially Disadvantaged Operators' Enrollment Trends. Economic Research Service, U.S. Department of Agriculture, EIB-62, 41, December.

Peltzman, S. & Tideman, T. N. (1972). "Local versus National Pollution Control: Note," *American Economic Review LXII:959-63*, December.

Pendell, Dustin, L., Jeffery, R. Williams, Scott, B. Boyles, Charles, W. Rice, & Richard Nelson, G. (2007). "Soil Carbon Sequestration Strategies with Alternative Tillage and Nitrogen Sources," *Review of Agricultural Economics, 29(2)*, (Summer):247-68.

Riley, T. Z. (2004). "Private-land habitat opportunities for prairie grouse through federal conservation programs," *Wildlife Society Bulletin, 32(1)*, 83-91.

Secchi, S., Gassman, P. W., Jha, M., Kurkalova, L., Feng, H. H., Campbell, T. & Kling. C. L. (2007). "The cost of cleaner water: Assessing agricultural pollution reduction at the watershed scale," *Journal of Soil and Water Conservation, 62(1)*, 10-21, January-February.

U.S. Department of Agriculture, Natural Resources Conservation Service. (2004b). Wildlife Habitat Incentives Program, Program Description. Dated September 2004, at http://www.nrcs.usda.gov/programs/farmbill/2002/ pdf/WHIPPrDs.pdf (last accessed September 4, 2008).

U.S. Department of Agriculture, Natural Resources Conservation Service. (2003). *Environmental Quality Incentives Program: A Benefit-Cost Analysis, Final Report*, Washington, DC. http://www.nrcs.usda.gov/ programs/Env _Assess/EQIP/EQIP _EA _finals _FINAL _BC _Analysis.pdf.

U.S. Department of Agriculture, Natural Resources Conservation Service. (2006). National Priorities, Programmatic, Fiscal Year 2006. Updated 1/24/2006 at http://www.nrcs.usda. ov/programs/natprgmpriorities/ FY2006/natprgmprior.html (last accessed September 4, 2008).

U.S. Department of the Interior, Fish and Wildlife Service, and U.S. Department of Commerce, Bureau of the Census. (1993). 1991 *National Survey of Fishing, Hunting, and Wildlife-associated Recreation*.

U.S. Department of the Interior, Fish and Wildlife Service, and U.S. Department of Commerce, Bureau of the Census. 1997. *1996 National Survey of Fishing, Hunting, and Wildlife-associated Recreation*.

U.S. Government Accountability Office (GAO). (2006). *USDA Should Improve Its Process for Allocating Funds to States for the Environmental Quality Incentives Program*. Report to the Ranking Democratic Member, Committee on Agriculture, Nutrition, and Forestry, U.S. Senate. GAO-06- 969, September.

End Notes

[1] Farm legislation refers to the FRPP as the Farmland Protection Program.

[2] Source of the 2002 Farm Act spending amounts is from the Office of Budget Policy and Analysis; 2008 Farm Act spending estimates are from the Congressional Budget Office.

[3] Of the working-land and land protection programs discussed here, only EQIP contains legislated provisions that target funding by producer type.

[4] The term "State" used in this chapter includes the Pacific Basin territories and Puerto Rico.

[5] The 2002 Farm Act allowed States to contribute up to 90 percent of the cost of structural practices (e.g., fences, lagoons, and vegetative buffers) incurred by beginning or limited-resource producers.

[6] Nearly a third of EQIP program funding is used for technical assistance to assist producers with conservation practice implementation. These funds have typically been allocated to States in proportion to the amount of financial assistance the States were awarded.

[7] NRCS reserves a small portion of funds for special reasons, such as performance incentive awards.

[8] The CRP does place a 25-percent cap on cropland enrolled in any one county, but there is no explicit enrollment cap at the State level.

[9] To implement the requirements of the Regional Equity provision, during 2004-06, NRCS took several steps: NRCS estimated which States were likely to be below the threshold, and by how much; NRCS withheld an amount equal to this expected shortfall from program budgets and allocated the remainder to States using the programs' allocation criteria; and NRCS then reallocated withheld funds to the underfunded States according to need.

[10] The choice of which conserva-tion programs are used to satisfy the Regional Equity requirement can affect the ultimate impact of the provision on program outcomes. ERS analysis (not reported here) revealed that it is possible to implement the Regional Equity provision by reallocating funds solely within EQIP because its budget is much larger than the other programs. Doing so would result in States' receiving combined funding levels (from all four programs) that are not strikingly different from the current method. However, because States would receive different amounts through each individual program and because each program targets different environmental problems at different costs, such a strategy would likely result in different environmental benefit and cost tradeoffs.

[11] Even in programs like EQIP that seek to maximize benefits, many States consider cost in enrollment decisions—so competition for limited funding could induce producers to choose to implement less costly conservation practices.

[12] The coefficient of variation (CV) is above 1.45 for both RE and non-RE States in each year. The CV is a measure of the variation in a population, and values exceeding one are typically considered high-variance.

[13] In some cases, the same acres received treatment for more than one resource concern, so adding up the net gains and losses in acres across categories would likely overstate the net effect of the Regional Equity reallocations on the total number of acres treated.

[14] Another important consideration was technical assistance costs (TA). The contract data contain information on the financial assistance (FA) costs of implementing practices, but do not identify which practices or resource issues required TA. TA costs are not insignificant outlays—in EQIP between 2004 and 2006, TA costs represented about 21 percent, 24 percent, and 28 percent of the *total* program budgets, respectively. In WHIP, TA costs were 22 percent and 28 percent in 2005 and 2006, respectively. We assumed that TA was necessary for meeting nondegradation standards, so in developing the per acre cost estimates we increased the costs to reflect TA based on these overall portions of the budget (e.g., in EQIP in 2006, TA represented an overall 40 percent of FA costs, so 40 percent was added to the cost of contracts in that year).

[15] The one exception was that, in WHIP, the marginal contracts in RE States provided sheet and rill erosion reduction benefits at a lower cost per acre.

[16] Conservation programs fund activities that can increase the amenity, scenic, and recreational value of agricultural land in ways that are not easily expressed in dollar terms. These values can be higher when more people have access to the amenities (Feather et al., 1999), which suggests that shifting program activity to higher population States can increase these nonmarket benefits.

[17] Farm and Ranch Lands Protection Program, Final Rule, Federal Register, Vol. 68, no 95, Page 26461-26478, May 16, 2003. Online at http://edocket.access

[18] Grassland Reserve Program Final Rule, Federal Register, Vol 69, no 99, Page 2918 1-29182, May 21, 2004. Online at http://frwebgate1 .access 418564507624+5+2+0&WAISaction= retrieve.

[19] Ranking scores reflect the agency's numerical determination of the priority for enrolling a particular contract with funds from a particular pool, and higher ranking scores imply higher priority. In a few instances, groups of contracts were assigned to one of three priority categories (priority = 1st, 2nd, or 3rd). We assigned ranks to the lowest priority contracts based on costs, consistent with the ranking approach in Kansas and Mississippi (the two States with the most contracts ranked on a priority basis). When a contract was missing a ranking score, we used regression analysis to predict the ranking score based on contract characteristics, but because

the decision and ranking processes are so decentralized, this effort was not successful even at a State level. Thus, we assigned the average ranking score from contracts within the relevant funding pool to unranked contracts.

[20] In this chapter we define structural practices as practices with a 15-year life that are likely to treat multiple fields (lagoons, ponds, etc). Vegetative practices include vegetative filter strips, buffers, and small structural practices that are likely to treat single tracts/fields. Management practices include, for example, integrated pest management, nutrient management, and conservation tillage.

[21] This method puts greater reliance on the NRCS field office judgments about which resource concerns a practice is actually intended to address. An alternative method, which might ensure more consistency across years, would be to impose a strict mapping between practices and the resource concerns they are intended to address. However, this alternative would limit the spatial variability in the data as well as the ability of practices to address multiple resource concerns. The spatial variability in this study is particularly important because we are distinguishing effects of the Regional Equity provision among different States.

[22] We define vegetative practices to include small structural practices that are likely to treat particular fields or tracts, as opposed to large structural practices that are more likely to treat physical effects on all fields.

In: U.S. Farms and Conservation Programs
Editor: Lucas A. Savarese
ISBN: 978-1-61209-927-9
© 2011 Nova Science Publishers, Inc.

Chapter 4

CONSERVATION RESERVE PROGRAM: STATUS AND CURRENT ISSUES

Tadlock Cowan

SUMMARY

The Conservation Reserve Program (CRP), enacted in 1985, provides payments to farmers to take highly erodible or environmentally sensitive cropland out of production for 10 years or more. It is the federal government's largest private land retirement program. The program is administered by the Farm Service Agency (FSA) of the U.S. Department of Agriculture (USDA), with technical assistance provided by USDA's Natural Resources Conservation Service. The CRP also has several subprograms, the best-known of which is the Conservation Reserve Enhancement Program (CREP).

The 2008 farm bill (P.L. 110-246) reauthorized CRP through FY2012 but reduced the maximum acreage level to 32 million acres, down from the previous cap of 39.2 million acres. Criteria for haying and grazing on CRP land were amended, and incentives were authorized to assist socially disadvantaged and beginning farmers in leasing or purchasing land under a CRP contract. A draft supplemental environmental impact statement (SEIS) on the 2008 farm bill changes to CRP was completed in February 2010. On May 14, 2010, an interim rule was published in the *Federal Register* to implement the Transition Incentives Program. The final SEIS was available for public comment for a 30-day period beginning June 18, 2010, and ending July 19, 2010. Following the end of the comment period, and a review of the comments by FSA, a record of decision (ROD) for the CRP SEIS will be issued.

The national enrollment as of July 2010 stood at 31.3 million acres, a decrease of approximately 3.4 million acres from October 2008 and a decrease of approximately 2.4 million acres from September 2009. Approximately 4.3 million acres of farmland was added through a general sign- up (number 39) in summer 2010, nearly 57% of which was acreage under contracts set to expire September 30, 2010. This was the first general sign-up since 2006. Approximately 85% of total CRP acreage is currently enrolled under general sign-ups.

There was also a continuous enrollment sign-up during late spring and summer 2009 (number 37) that added 488,000 acres to CRP totals. Continuous sign-up 38 began in October 2009 and will continue to October 2011.

For FY2010, rental outlays are projected to total $1.9 billion, approximately $250 million less than for FY2009. This projected total includes funding for rental payments, cost-share payments, and incentive payments. The average per-acre rental payment for general sign-ups is currently $44.57, and the average rental rate for CREP totals over $129 per acre. The average rental payment for all CRP programs is approximately $53 per acre.

Between 2007 and 2010, 27.8 million acres under CRP contracts expired. Contracts for approximately 24 million (86%) of these acres have been renewed or extended. On September 30, 2009, contracts on approximately 3.9 million acres were set to expire. USDA announced a sign- up for contract extensions that ran from May 18 to June 30, 2009. Of the expiring 3.9 million acres, however, only about 1.5 million were offered extension contracts. About 55% of the eligible expiring acreage was in four states: Colorado, Kansas, Montana, and Texas. As of December 2009, participants holding contracts on 1.1 million acres originally set to expire September 30, 2009, had accepted extension offers (73%). Contracts on an additional 4.5 million acres are set to expire September 30, 2010. Approximately 75% of the FY20 10 expiring acreage has been reenrolled or had its contracts extended.

CRP General Background

The Conservation Reserve Program (CRP) is the federal government's largest land retirement program for private land. It was first enacted by Congress in 1985 to help control soil erosion, stabilize land prices, and control excessive agricultural production. Since then, program purposes have been expanded to include environmental goals. The program is administered by USDA's Farm Services Agency (FSA), with technical assistance from USDA's Natural Resources Conservation Service (NRCS) and funding from USDA's Commodity Credit Corporation (CCC). The FSA makes annual rental payments based on the agriculture rental value of the land, and provides cost-share assistance for up to 50% of the participant's costs in establishing various approved conservation practices. There are also one-time sign-up bonuses and incentives for socially disadvantaged farmers and ranchers, new and beginning famers and ranchers, and limited-resource farmers and ranchers. Participants enroll in CRP contracts for 10 to 15 years.

Participants bid to retire land from production for 10-15 years. Contracts are awarded by FSA based on their assessment of the land's environmental value using an Environmental Benefits Index (EBI). If the land is accepted, the landowner may enroll the land, receive annual rental payments for it, and maintain the land under an approved conservation plan. After a CRP contract expires, federal payments cease. If the land in question is "highly erodible" (about 75% of the land enrolled in the CRP meets this definition) and participants decide to return the land to production, they must manage this land under a NRCS-approved conservation system to be eligible for some federal farm programs (including commodity payments).

ENROLLING IN CRP

There are two types of sign-ups for enrolling land in the CRP: general and continuous. As of July 2010, there were 743,449 CRP contracts nationally on 416,121 farms (nearly 19% of all farms) under all CRP programs. These data compare to 788,118 CRP contracts on 443,615 farms in September 2007. Contracts under continuous sign-ups now exceed the number of contracts under general sign-ups (401,117 versus 342,332).[1]

Land eligible for CRP enrollment must be either (1) cropland that is planted or considered planted to an agricultural commodity in four of the previous six crop years from 1996 to 2001, or (2) certain marginal pastureland that is enrolled in the Water Bank Program or suitable for use as a riparian buffer or for similar water quality purposes. The 2008 farm bill modified the land eligibility requirements so that alfalfa and other multi-year grasses and legumes are to be considered agricultural commodities when grown in a rotation practice approved by FSA. The farm bill also clarified that when alfalfa is grown in an approved rotation practice, it is to be considered an agricultural commodity and can be used to fulfill the requirement that land be cropped in four of six previous years in order to be eligible.

General Sign-Up

General sign-ups are specified enrollment periods during which landowners compete nationally to enroll their land in the CRP. A new general sign-up (number 39) was held in late spring and summer 2010. This was the first general sign-up since 2006. For this most recent general sign-up, FSA accepted 4.3 million acres. Over half the acreage enrolled was set to expire on September 30, 2010. Approximately 85% of CRP acreage (26.7 million of 31.3 million acres) is currently enrolled through general sign-up. Applicants must meet certain eligibility criteria, evaluate their land according to FSA's Environmental Benefits Index, and submit bids to FSA for enrollment. FSA accepts applications that demonstrate the highest environmental benefits.

These general sign-ups are always competitive. For the most recent sign-up, nearly 4.8 million acres were offered for sign-up, of which 4.3 million were accepted. For the previous general sign- up (number 33), which ran from March 27 to April 28, 2006, USDA selected 1 million acres of the 1.4 million acres offered. This enrolled acreage included about 673,000 acres of land located within conservation priority areas, about 629,000 acres with an erodibility index of eight or greater (highly erodible), and about 265,000 acres to be restored to rare and declining habitats.

Environmental Benefits Index (EBI)

As the CRP has been expanded to include broader environmental goals, FSA has adjusted the categories and points awarded under the EBI. For example, FSA announced in June 2003 that, for the first time, it could award points to projects that have the potential to sequester carbon (reducing greenhouse gas emissions). Other factors include wildlife habitat benefits from planted cover crops, water quality benefits from reduced erosion, and whether benefits will endure beyond the contract period. Offers that included a willingness to accept less than the maximum rental rate for acreage, as well as offers to reenroll land, may have received additional points. FSA ranks all applications nationally, and then sets an EBI score cutoff above which applications will be accepted. For the most recent general sign-up, FSA set a cutoff score of 200 on the EBI.

Continuous Sign-Up (Includes Bobwhite Quail and Non-Floodplain Wetlands)

Environmentally desirable land devoted to specific conservation practices with high environmental benefits may be enrolled in the CRP at any time for 10-15 years under continuous sign-up.[2] Offers are automatically accepted (provided the land and producer meet certain eligibility requirements) and are not subject to competitive bidding. Contracts usually include additional incentive payments. Within the continuous sign-up program there are some options tailored to certain conservation needs, such as restoring floodplain wetlands and native hardwood trees in wetlands. In August 2004, the Administration announced two more initiatives: a 25 0,000- acre initiative to restore bobwhite quail habitats in the Midwest and the Southeast, and a 250,000- acre initiative to restore wetlands located outside floodplains (including Great Plains playa lakes). The 2008 continuous sign-up (number 36) added 282,000 acres, the continuous sign-up for FY2009 (number 37) added 277,556 acres as of July 2009, and the FY20 10 sign-up (number 38) has added 296,615 acres as of July 2010. These sign-ups bring the total acreage under continuous sign-up contracts to over 4.6 million acres as of July 2010, about 15% of the total 31.3 million acres CRP acres.[3]

Conservation Reserve Enhancement Program (CREP)

CREP is a joint federal-state continuous sign-up program. Currently there are 43 agreements in 32 states. CREP targets geographic areas with agriculture-related environmental problems, such as Maryland's Chesapeake Bay and Florida's Everglades. Some states (e.g., New York and Ohio) have multiple CREPs, each targeting a different area of the state. USDA provides 80% of the funding, and a non-federal entity (typically the state) contributes the remainder. States may automatically enroll up to 100,000 acres. Unlike the general sign-up, CREP both encourages landscape-scale conservation efforts and offers the flexibility to address locally identified needs. As of July 2010, nearly 1.2 million acres were enrolled in CREP, over 100,000 acres more than in September 2007.[4]

Farmable Wetlands Program (FWP)

As authorized under the 2002 farm bill, farmable wetlands—those wetlands that have been partially drained, or are naturally dry enough to allow crop production in some years, but otherwise meet the definition of a wetland—may be enrolled in the CRP on a continuous basis. Up to 100,000 acres may be enrolled from any state (this may be increased to 150,000 acres after three years). The farm bill reserved one million acres for farmable wetlands enrollment. As of July 2010, 230,948 acres were enrolled in the programs.

The 2008 farm bill expanded land eligibility for FWP from land that was cropped during at least 3 of the immediately preceding 10 crop years, as well as contiguous buffer acreage to protect the wetlands, to include:

- land on which constructed wetland is to be developed that will receive flow from a row-crop agriculture drainage system and is designed to provide nitrogen removal in addition to other wetland functions;
- land devoted to commercial pond-raised aquaculture in any year during calendar years 2002-2007;
- intermittently flooded land, provided the land had a cropping history in three years between 1990 and 2002 and was subject to natural overflow of prairie wetland; and
- buffer acreage that enhances wildlife benefits. On a single tract of land, enrollment is now set at a maximum of 40 contiguous wetland acres. "Flooded farmland" has a 20-acre limit. No commercial use of the land is permitted.

Other CRP Programs

Several other CRP initiatives support additional specific conservation practices. These include acreage in upland bird habitat buffers, bottomland hardwood trees, non-flood plain and playa wetlands, longleaf pine plantings, the Prairie Pothole duck nesting habitat, and state acres for wildlife enhancement. In February 2010, Secretary Vilsack announced the expansion of acreage allocated to three of these programs: state acres for wildlife enhancement was increased by 150,000 acres; upland bird habitat buffers were increased by 100,000 acres; and duck nesting habitat was increased by 50,000 acres. **Table 1** provides acreage data for these programs as of July 2010 and cumulative acreage totals for each program.

Table 1. CRP Initiatives and Acreage Enrollment (July 2010).

CRP Initiative and Acreage Goal	Acreage Year-to-Date	Cumulative Acreage[a]
Farmable wetland program (up to 1 million acres)	25,014	296,615
Upland bird habitat buffers (350,000 acres)	11,496	228,377
Bottomland hardwood trees (500,000 acres)	6,894	59,152
Non-flood plain and playa wetlands (250,000 acres)	20,368	98,411
Flood-plain wetlands (500,000 acres)	23,200	175,037
Long-leaf pine plantings (250,000 acres)	8,165	86,342
Prairie Pothole duck nesting habitat (150,000 acres)	15,883	111,785
State acres for wildlife enhancement (650,000 acres)	100,965	357,367

Source: Farm Service Agency.

[a] Includes acreage under sign-up 38 contracts that begin in FY20 11.

CURRENT ISSUES

CRP Provisions in the 2008 Farm Bill

The 2008 farm bill (P.L. 110-246) reauthorized the CRP with a maximum acreage cap of 32 million acres, down from a cap of 39.2 million acres established in the 2002 farm bill (P.L. 107- 171). The farm bill further directed the Secretary of Agriculture to give priority (where environmental benefits are equal) in contract bids to producers who reside in the county where the CRP acreage is located. FY2009 enrollment ended at 33.7 million acres. Following expiration of 2.8 million acres, FY2010 began with 30.93 million acres under contract. Current total enrollment is 31.3 million acres, approximately 723,000 acres below the cap.

The 2008 farm bill also amended the pilot program for wetland and buffer acres in CRP. Each state can enroll up to 100,000 acres, up to a national maximum of 1 million acres. This maximum may be raised to 200,000 in each state following a review of the program. Eligible lands for the program include (1) wetlands that have been cropped in three of the immediately preceding 10 crop years; (2) land on which a constructed wetland is to be developed to manage fertilizer runoff; and (3) land that has been devoted to commercial pond-raised aquaculture.

The farm bill permits 50% cost-share payments on land used for hardwood trees, windbreaks, shelterbelts, and wildlife corridors for contracts entered into after November 1990. Contracts extend from a minimum of two years up to four years. Funding of $100 million also was authorized to cover cost-sharing for the thinning of trees to improve the management of natural resources on the land.

The 2008 farm bill also modified the criteria for evaluating CRP contract applications. Evaluation criteria include the extent to which a CRP contract application would improve soil resources, water quality, or wildlife habitat. The bill also allowed the Secretary to establish different criteria in various states or regions that lead to improvements in soil quality or wildlife habitat. Preference in new CRP contracts will be given to land owners and operators who are residents of the county or a contiguous county in which the land is located.

Other CRP provisions in the 2008 farm bill included a redefinition of the Chesapeake Bay region as a priority area without limiting the region to the states of Pennsylvania, Maryland, and Virginia. While the new program applies to all watersheds draining into the Chesapeake Bay, the Susquehanna, Shenandoah, Potomac, and Patuxent Rivers get funding priority.

The 2008 farm bill also established incentives to increase the participation of beginning and socially disadvantaged farmers and ranchers.[5] The CRP Transition Option is designed to facilitate the transition of land to beginning and socially disadvantaged farmers and ranchers for the purpose of returning the land to production using sustainable grazing or crop production methods. To encourage this program, CRP contract holders can receive two extra years of rental payments for leasing or selling their land to a beginning or socially disadvantaged farmer. The National Organic Certification Cost Share Program can be used in conjunction with the CRP Transition Option to defray the costs, up to $750 per year, of organic certification for producers and handlers of organic produce. It authorizes CRP contract modifications to assist these producers in leasing or purchasing land under a CRP contract from a retired or retiring farm owner or operator. The provision authorizes $25 million for assistance in making these land transfers.

The CRP Transition Option works as follows:

- One year prior to the termination of a CRP contract, a CRP owner or operator who is participating in the CRP Transition Option can join with a beginning or socially disadvantaged farmer or rancher who can begin to make conservation and land improvements and/or begin the organic certification process on the land covered by the CRP contract.
- On or near the date that the CRP contract is terminated, the retired or retiring owner or operator will either sell some or all of the land that was covered by CRP to the participating beginning or socially disadvantaged farmer or rancher, or enter into a long-term lease with the farmer/rancher, or lease with an option to purchase.
- The participating beginning or socially disadvantaged farmer or rancher must develop and implement a conservation plan on the land that was covered by CRP.
- On the date that the participating beginning or socially disadvantaged farmer or rancher takes possession of the land through ownership or lease, the farmer/rancher will have the option to enroll in the Conservation Stewardship Program or the Environmental Quality Incentives Program. The farmer/rancher will also have the option of re-enrolling portions of the land into the CRP through the continuous sign-up CRP.
- FSA will continue making payments to the retired or retiring owner or operator for two additional years after the date that the CRP contract terminates.

Farm Bill Implementation Issues

FSA divided CRP-related changes into two regulations: Part 1, those requiring an environmental impact statement (EIS), and Part 2, those requiring only an environmental assessment (EA). Both the EIS and EA discuss potential environmental impacts and alternative courses of action, although the EA is a more limited analysis. The EIS process can take 18-24 months, or longer in some cases, while an EA can often be completed within a year or less.

In June 2009, FSA issued a rule implementing all changes that fell under Part 2, the EA-associated changes.[6] These include an assessment of the three new practices under the Farmable Wetlands Program: (1) eligibility for commercial pond-raised aquaculture; (2) flooded prairie farmland; and (3) constructed wetlands. Other EA-associated changes are authorization of $50 million cost-share assistance for tree trimming, refining income limits for program eligibility, and the imposition of the 32 million-acre cap. Each of these changes were included in the Part 2 rule promulgated in June 2009.

The Part 2 CRP regulation will implement provisions that require an EIS. These provisions include:

- updating crop history eligibility to include four of the last six years between 2002 and 2007;

- exempting certain CREP and continuous CRP acres from the county enrollment cap of 25% of cropland;
- implementing the new transition incentives for beginning and socially disadvantaged farmers and ranchers programs; and
- routine grazing policy on CRP land.

The CRP Transition Option program could be delayed by the EIS process. FSA published the draft supplemental environmental impact statement (SEIS) in February 2010.[7] The final SEIS was available for public comment for a 30-day period beginning June 18, 2010, and ending July 19, 2010.[8] Following the end of the comment period, and a review of the comments by FSA, a record of decision (ROD) for the CRP SEIS will be issued.

USDA's justification for the environmental study is that while the land is leaving the conservation program, the beginning farmers would be allowed to start working the land a year before the contract is up. Landowners normally are barred from farming or grazing the land until the contract expires. The Sustainable Agriculture Coalition and the Center for Rural Affairs each ran letter campaigns in 2009 urging FSA to implement the program without conducting an EIS. Other groups wrote to Secretary Vilsack requesting that implementation of the CRP Transition Option not be delayed by an environmental assessment. Wildlife conservation groups (e.g., the National Wildlife Federation) supported the decision to prepare the EIS. In comments on the draft SEIS, the National Grain and Feed Association has requested the FSA to lower the acreage enrolled in CRP to 24 million acres and to target conservation spending on working farmland (e.g., Environmental Quality Incentives Program, Conservation Stewardship Program).

On May 14, 2010, an interim rule was published in the *Federal Register* to implement the Transition Incentives Program.[9] The rule essentially amends the CRP contract regulations to provide that retired or retiring farm owners and operators have permission to amend their contract if it is due to expire within one year to facilitate the transition of the land enrolled in the expiring CRP contract to a beginning or socially disadvantaged farmer or rancher. Acquiring farmers may return some or all of the land into production provided they use sustainable grazing or crop production methods. The 2008 farm bill authorizes $25 million for incentive payments to retired or retiring owners and operators who sell land or lease it long-term to beginning and socially disadvantaged farmers and ranchers.

Managed Haying and Grazing

Managed haying and grazing are permitted activities under certain conditions. A disaster declaration, for example, may permit contract holders to harvest hay or to graze their cattle for certain periods. Where such activity is permitted, contract payments may be reduced. The 2008 farm bill modifies the managed haying and grazing regulation (Section 2108) to permit routine grazing to control invasive species under specific conditions and also permits the installation of wind turbines under certain conditions. Where routine harvesting is permitted, state technical committees are required to coordinate to ensure appropriate environmental management. When such activities are permitted, rental payments will be reduced by an amount commensurate with the economic value of the authorized activity. The farm bill also prescribed grazing for control of invasive species as a permissible activity. The managed grazing and haying and installation of wind turbines are assessed in the SEIS.

Haying and Grazing for Critical Feed Use (CFU)

In May 2008, USDA announced that 24 million acres of CRP land could be used in 2008 for a critical feed use (CFU) program of managed haying and grazing following primary bird nesting season. This contract modification is restricted to the least environmentally sensitive land and will limit the scope and frequency of any managed haying and grazing. The National Wildlife Federation (NWF) and six state NWF chapters sought an injunction against USDA for failure to conduct an appropriate environmental review of the proposed CFU program. The U.S. District Court for the Western District of Washington issued a temporary restraining order (TRO). The court then issued a permanent injunction suspending the CFU provision except for those who had been approved by or had applied to FSA prior to the TRO, or who had invested at least $4,500 toward haying or grazing equipment and preparation prior to the TRO. Where the application was approved prior to the TRO, haying and grazing had to be completed by November 10, 2008. The CFU exemption was not authorized for FY2009, and there are no current plans to authorize a CFU program for FY20 10.

Expiring CRP Contracts and Reenrollment and Extension Policy

Approximately 28 million acres under CRP contract will expire between 2007 and 2010. Contracts covering 5.9 million acres were set to expire in FY2008, and 3.9 million more acres were set to expire at the end of FY2009. An additional 4.5 million acres will also expire September 30, 2010. Approximately 85% of the acreage expiring between 2007 and 2010 has been reenrolled or has had contracts extended as of July 2010, including 44,201 acres approved for enrollment in FY20 11. Of the 15.7 million contract acres that expired September 30, 2007, 13.6 million (86.6%) were approved for extensions or new enrollment contracts. Approximately 4.8 million of the 5.9 million acres that expired in 2008 also have extensions or new contracts (81.6%). For contracts expiring September 30, 2010, about 75% of the 4.5 million acres has been reenrolled or had contracts extended.

Contracts were extended or renewed based on the Environmental Benefit Index (EBI) score and the land's location within national priority areas.[10] FSA ranked individual contracts into one of five tiers based on the environmental benefits of the original EBI score. Eligible participants ranking in the first tier (81%-1 00% of the EBI) could reenroll their land in new 10-year contracts. Farmers and ranchers in this top tier with wetlands enrolled were eligible for 15-year contracts. Only acreage under general sign-up contracts is eligible. Eligible participants ranking in the second tier (61 %-80% of the EBI) could extend their contracts for five years. Third-tier participants (21 %-40% of the EBI) could receive three-year extensions. Eligible participants in the bottom tier could extend their expiring contracts by two years.

USDA announced a sign-up for voluntary extensions between May 18 and June 30, 2009. Of the expiring 3.9 million acres for FY2009, approximately 100,000 acres were under continuous sign- up and were reenrolled if the producer wished to do so. However, only about 1.5 million acres enrolled under the general sign-up were offered contract extensions. Of these, 1.1 million acres were reenrolled as of December 2009 (73%). Acreage eligible for contract extensions had to score in the top 30% of the EBI or have an Erodibility Index (EI) of 15 or higher. Participants who were eligible for contract extensions were notified by letter from their county Farm Service Agency office. The reduction in contract extensions was necessary to ensure that the total acreage under contract would conform to the new 32 million-acre limit by the end of FY2009. Contract extensions ranged from two to five years at the same per-acre payment rate. Colorado, Kansas, Montana, and Texas had the largest number of eligible acres with a 30% EBI or a ranking of 15 on the EI. These four states accounted for 55% of the total national acres eligible for extension in FY2009.

PROGRAM COSTS AND BENEFITS

Acres enrolled in CREP, continuous enrollments, or the farmable wetlands program are generally eligible for higher payments than acres enrolled under general sign-ups because of their higher environmental benefits, location and prevailing rental rates, and additional financial incentives for participation. However, such contracts involve much smaller acreage on average. CREP payments average over $129 per acre and $116 for the FWP, versus an average per-acre payment of approximately $44 for the general sign-up acreage.

FSA estimates approximately $1.7 billion in projected outlays for CRP payments under all programs in FY2010. This is about $250 million less than for FY2009. This projected total includes funding for rental payments, cost-share payments, and incentive payments. The Congressional Budget Office had estimated that CRP contract obligations would cost approximately $2.4 billion annually through 2017. This was before the reduction in acreage to 32 million acres. NRCS estimated that, prior to 2003, monetized CRP benefits (such as increased wildlife habitat and small game hunting) averaged about $1.4 billion per year. This figure does not include non-monetized benefits such as improved groundwater quality and wetland restoration.

Rental Rates for CRP Acreage

The average rental rate for all CRP land was $53.24 per acre as of July 2010. Rental rates range from an average of $44 for general sign-up acreage to over $129 for CREP acreage. CRP rental rates are based on the three-year average of local dry-land cash rental rates. An up-front signing incentive payment (SIP) of $100 to $150 per acre (depending on contract length) is available for eligible participants who enroll certain practices. The one-time SIP is made after a contract is approved and all payment eligibility criteria have been met. A practice incentive payment (PIP) equal to 40% of the eligible installation costs is also available for eligible participants who enroll certain practices on their acreage.

Rental rates for CRP contracts became an important issue to some producers when commodity prices rose in 2008. The producers claimed that CRP rental rates were significantly lower than the producers could get by renting their land out for production. Many producers pressured USDA for penalty-free contract terminations. If rental rates are too low, there is some chance that producers will decline to enroll their land, or, if enrolled, will decline to renew their contracts at expiration. Putting CRP acreage back into production could have significant environmental effects. Although land put back into production would have to be managed under an environmental plan to be eligible for various agricultural assistance programs, there could still be an environmental cost in terms of increased sediment losses and nitrogen and phosphorus run-off if fragile land were put back into production.[11] The decline in commodity prices from their 2008 highs reduced the pressures on producers to terminate their contracts. The 2008 farm bill (Section 2110) directs USDA to conduct an annual survey of per-acre estimates of the average market dry land and irrigated land cash rental rates and to post these rates on a publicly accessible USDA website.

Contract Termination and Penalty Fees

High grain prices in the first half of 2008 appeared to make contract terminations attractive to some producers. Under current law, however, a producer wishing to terminate a contract early faces a penalty fee of 25% on rental payments paid, plus repayment, with interest, of all the funds already paid to the producer. This includes any cost-share payments. CRP acreage is also seen in some quarters as a potential resource for renewable fuel feedstocks.

Tax Status of CRP Payments

CRP rental payments are regarded by the Internal Revenue Service (IRS) as income from the business of farming. As such, they are subject to self-employment Social Security taxes. Producers, however, would like to treat CRP payments as rental income not subject to the self- employment tax of 15.3%. The IRS position was supported by the Sixth Circuit Court in March 2000 in *Wuebker v. Commissioner*, 205 F.3d897. In December 2006, the IRS issued Notice 2006- 108 reinforcing its position that CRP payments are subject to self-employment taxes. Section 15301 of the 2008 farm bill exempts CRP payments from self-employment taxes for disabled and retired contract holders after December 31, 2007.[12]

Effects of CRP on Local Economies

Retiring land in rural, largely agricultural economies could result in fewer farmers and fewer farming-related jobs in these areas. A USDA report found that, although high CRP enrollment was associated with some job loss in rural areas between 1986 and 1992—the years the CRP was first underway—this was generally not the case during the 1990s. However, the report noted that national trends could mask regional adjustments, and that "local economic adjustments might be sizeable."[13] Losing existing CRP acreage or halting new enrollments may also have effects on local economies where hunting and fishing are important economic activities.

By statute, CRP enrollment is capped at 25% of a county's cropland. According to FSA, approximately 130 counties have at least 22.5% enrolled, although this can include counties with very small total acreage of cropland. Of these 130 counties, 80 have at least 25%

enrolled. The farm bill exempts CREP and continuous sign-up acreage from the 25% county cap.

Some groups believe that retiring land from productive agricultural use is detrimental. In comments on the SEIS noted above, the president of the National Grain and Feed Association argued that CRP has substantial negative economic impacts on agricultural production and rural communities:

> If U.S. agriculture is to achieve economic growth, which in turn contributes to job creation and revitalization of rural communities, it is imperative that the CRP be right-sized and reformed. The idling of productive land resources cuts off the economic multiplier inherent in crop, livestock and poultry production, thereby costing jobs and suffocating economic vitality in rural communities. Further, it risks compromising the United States' ability to provide abundant, affordable and cost-competitive food, animal feed, exports and biofuels.[14]

CRP Environmental Effects

FSA estimates that, compared with 1982 erosion rates, the CRP has reduced erosion by more than 454 million tons per year on the 34.6 million acres enrolled in the program. Through April 2006, CRP had also restored 2 million acres of wetlands and 2.5 million acres of buffers. Other conservation benefits NRCS has documented on these lands include the sequestration of more than 48 million metric tons of carbon annually; more than 3.2 million acres of wildlife habitat established; and a reduction in the application of nitrogen (by 681,000 tons) and phosphorus (by 104,000 tons). Also, participants have planted about 2.7 million acres of trees, making it the largest federal tree-planting program in history.

End Notes

[1] The five states with the largest number of acres under CRP contract are Colorado, Kansas, Montana, North Dakota, and Texas. These five states account for about 44% of the total national acreage enrolled.

[2] Specific conservation practices include filter strips, riparian buffers, grass waterways, shelterbelts, field windbreaks, living snow fences, salt-tolerant vegetation, shallow water areas for wildlife, wetland restoration, and wellhead protection areas.

[3] This figure excludes 83,690 acres under sign-up 38 that start in FY20 11.

[4] This figure excludes approximately 44,000 acres under sign-up 38 that start in FY20 11.

[5] Section 2111 of the Food, Conservation, and Energy Act of 2008 amends Section 123 5(c)(1)(B) of the Food Security Act of 1980, codified at 16 U.S.C. § 3835(c)(1)(B), to create the Conservation Reserve Program Transition Incentives for Beginning and Socially Disadvantaged Farmers and Ranchers.

[6] The rule may be accessed at http://frwebgate3.access 4468519271+40+1+0&WAISaction=retrieve.

[7] The draft SEIS can be accessed at http://www.agri-pulse.com/uploaded/201002CRPEIS.pdf.

[8] The final SEIS can be accessed at http://www.fsa.usda.gov/Internet/FSA_File/crpfinalseismaster61010.pdf.

[9] *Federal Register*, vol. 75, no. 93, May 14, 2010, at http://frwebgate4.access WAISdocID=38 153123 842+0+2+0&WAISaction=retrieve.

[10] National Priority Areas named in CRP authorizing legislation are the Chesapeake Bay, Long Island Sound, and the Great Lakes Region. USDA established two other national priority areas: Prairie Pothole Region in the Northern Great Plains, and Longleaf Pine Region in the southeast.

[11] For an examination of the potential environmental costs in Iowa as projected corn prices rose, see Silvia Secchi and Bruce Babcock, "Impact of High Corn Prices on CRP Acreage," *Iowa Ag Review*, Center for Agriculture and Rural Development, Iowa State University, Spring , 2007, at http://www.card.iastate.edu/iowa _ag_review/spring_07/ article2.aspx.

[12] For more detail on CRP's tax-related issues, see CRS Report RS22910, *The 2008 Farm Bill: Analysis of Tax-Related Conservation Reserve Program Proposals*, by Carol A. Pettit

[13] U.S. Department of Agriculture, Economic Research Service Report to Congress. *The Conservation Reserve Program: Economic Implications for Rural America.* September 2004.

[14] "NGFA Urges USDA to Downsize CRP to Protect Jobs & Growth," *Agri-Pulse*, April 13, 2010, at http://www.agripulse.com/20100413H1.asp.

CHAPTER SOURCES

The following chapters have been previously published:

Chapter 1 – This is an edited, excerpted and augmented edition of a United States Department of Agriculture Economic Research Report Number 44, dated October 2007.

Chapter 2 – This is an edited, excerpted and augmented edition of a United States Department of Agriculture Economic Information Bulletin Number 62, dated December 2009.

Chapter 3 – This is an edited, excerpted and augmented edition of a United States Department of Agriculture Economic Research Report Number 98, dated June 2010.

Chapter 4 – This is an edited, excerpted and augmented edition of a United States Congressional Research Service publication, Report Code RS21613, dated September 15, 2010.

INDEX

A

B

C

L

M

N

O

P

U

V

W

Y